D0942026

ASE Test Preparation Series

Truck Equipment

Auxiliary Power Systems Installation and Repair (Test E3)

DELMAR
CENGAGE Learning·

Australia • Brazil • Japan • Korea • Mexico • Singapore • Spain • United Kingdom • United States

DELMAR
CENGAGE Learning·

Delmar Learning's ASE Test Preparation Series: Truck Equipment Test Series Auxiliary Power Systems Installation and Repair (Test E3)

Vice President, Technology and Trades Professional Business Unit: Gregory L. Clayton

Director, Professional Transportation Industry Training Solutions: Kristen L. Davis

Editorial Assistant: Danielle Filippone

Development Editor: Dawn Jacobson

Director of Marketing: Beth A. Lutz

Marketing Manager: Jennifer Barbic

Senior Production Director: Wendy Troeger

Production Manager: Sherondra Thedford

Content Project Management: PreMediaGlobal

For product information and technology assistance, contact us at **Cengage Learning Customer & Sales Support, 1-800-354-9706**

For permission to use material from this text or product, submit all requests online at **www.cengage.com/permissions.** Further permissions questions can be e-mailed to **permissionrequest@cengage.com**

Library of Congress Control Number: 2011941744

ISBN-13: 978-1-4354-3937-5

ISBN-10: 1-4354-3937-6

Delmar Cengage Learning
5 Maxwell Drive
Clifton Park, NY 12065-2919
USA

Cengage Learning products are represented in Canada by Nelson Education, Ltd.

For more information on transportation titles available from Delmar, Cengage Learning, please visit our website at **www.trainingbay.com**

For more learning solutions, please visit our corporate website at **www.cengage.com**

Notice to the Reader
Publisher does not warrant or guarantee any of the products described herein or perform any independent analysis in connection with any of the product information contained herein. Publisher does not assume, and expressly disclaims, any obligation to obtain and include information other than that provided to it by the manufacturer. The reader is expressly warned to consider and adopt all safety precautions that might be indicated by the activities described herein and to avoid all potential hazards. By following the instructions contained herein, the reader willingly assumes all risks in connection with such instructions. The publisher makes no representations or warranties of any kind, including but not limited to, the warranties of fitness for particular purpose or merchantability, nor are any such representations implied with respect to the material set forth herein, and the publisher takes no responsibility with respect to such material. The publisher shall not be liable for any special, consequential, or exemplary damages resulting, in whole or part, from the readers' use of, or reliance upon, this material.

Printed in the United States of America
1 2 3 4 5 6 7 15 14 13 12 11

Contents

Section 6   Additional Test Questions for Practice

Section 7   Appendices

Glossary 87

Preface

Delmar, a part of Cengage Learning, is very pleased that you have chosen our ASE Test Preparation Series to prepare yourself for the Truck Equipment, Auxiliary Power Systems Installation Repair (E3) ASE examination. This guide is designed to introduce you to the task list for the Auxiliary Power Systems Installation Repair (E3) test you are preparing to take, give you an understanding of what you are expected to be able to do in each task, and take you through sample test questions formatted in the same way the ASE tests are structured.

If you have a basic working knowledge of the discipline you are testing for, you will find the Delmar Learning's ASE Test Preparation Series to be an excellent way to understand the "must know" items to pass the test. This book is not a textbook. Its objective is to prepare the technician who has the requisite experience and schooling to challenge ASE testing. It cannot replace the hands-on experience or the theoretical knowledge required by ASE to master vehicle repair technology. If you are unable to understand more than a few of the questions and their explanations in this book, it could be that you require either more shop-floor experience or further study.

This book begins with an item-by-item overview of the ASE Task List, with explanations of the minimum knowledge you must possess to answer questions related to the task. Following that, there are two sets of sample questions, followed by an answer key to each test and an explanation of the answers to each question. A few of the questions are not strictly ASE format but were included because they help teach a critical concept that will appear on the test. We suggest that you read the complete task list overview before taking the first sample test. After taking the first test, score yourself and read the explanation to any questions that you were not sure about, including the questions you answered correctly. Each test question has a reference back to the related task or tasks that it covers. This will help you to go back and read over any area of the task list that you are having trouble with. Once you are satisfied that you have all of your questions answered from the first sample test, take the additional test and check it. If you pass these tests, you will do well on the ASE test.

Our Commitment to Excellence

Delmar Cengage Learning has sought out the best technicians in the country to help with the development of this first edition of the Auxiliary Power Systems Installation Repair (E3) ASE Test Preparation Guide.

About the Author

Peter Taskovic has been in the automotive and truck repair industry and 2nd-stage manufacturing business for over 30 years. He has built and run everything from front-engine dragsters to 40-metric ton truck-mounted cranes. Peter has been ASE certified for over 25 years and currently holds Master Level Certificates in Automotive, Truck Repair, Body Refinishing, and Truck Equipment Repair as well as individual Alternative Fuels and Service Consultant certificates. He is also certified as a CSA Fuel System Inspector for CNG Vehicles. Peter played in integral part of the team that helped write the E Series Tests for ASE. Currently he holds the position of Manager of Technical Services at Auto Truck Group, LLC, where he has been employed for 25 years.

Thanks for choosing Delmar Learning ASE Test Preparation Series. All of the writers, editors, and Delmar staff have worked very hard to make this test preparation guide second to none. We know you are going to find this book accurate and easy to work with. It is our objective to constantly improve our products at Delmar by responding to feedback.

If you have any questions concerning the books in this series, please visit us on the Web at http://www.trainingbay.com

The History and Purpose of ASE

ASE began as the National Institute for Automotive Service Excellence (NIASE). It was founded as a nonprofit, independent entity in 1972 by a group of industry leaders with the single goal of providing a means for consumers to distinguish between incompetent and competent technicians. It accomplishes this goal by testing and certifying repair and service professionals. From this beginning, it has evolved to be known simply as ASE (Automotive Service Excellence) and today offers more than 40 certification exams in automotive, medium/heavy-duty truck, collision, engine machinist, school bus, transit bus, parts specialist, automobile service consultant, and other industry-related areas. There are now more than 400,000 professionals with current ASE certifications. These professionals are employed by new car and truck dealerships, independent garages, fleets, service stations, and franchised service facilities, to name a few. ASE continues its mission by providing information that helps consumers identify repair facilities that employ certified professionals, through its Blue Seal of Excellence Recognition Program. Shops that have a minimum of 75 percent of their repair technicians ASE certified and meet other criteria can apply for and receive the Blue Seal of Excellence Recognition from ASE.

ASE recognized that educational programs serving the service and repair industry also needed a way to be recognized as having the faculty, facilities, and equipment required to provide quality education to students wanting to become service professionals. Through the combined efforts of the ASE, the industry, and leaders in education, the nonprofit National Automotive Technicians Education Foundation (NATEF) was founded to evaluate and recognize training programs. Today, more than 2,000 programs are ASE certified under standards set by the service industry. ASEINATEF also has a certification-of-industry-(factory) training program known as Continuing Automotive Service Education (CASE). CASE recognizes training programs offered by replacement parts manufacturers as well as vehicle manufacturers.

ASE's certification testing is administered by American College Testing (ACT). Strict standards of security and supervision at the test centers ensure that the technician who holds the certification has earned it. Additionally, ASE certification requires the candidate to be able to demonstrate two years of work experience in the field before certification. Test questions are developed by industry experts who are actually working in the field. More details on how the test is developed and administered is provided in Section 2. Paper-and-pencil tests are administered twice a year at over 700 locations in the United States. Computer-based testing is now available, with the benefit of instant test results at certain established test centers. The certification is valid for five years and can be recertified by retesting. ASE issues a jacket patch, a certificate, and a wallet card to certified technicians and makes signs available to facilities that employ ASE-certified technicians. This is to enable consumers to recognize certified technicians.

You can contact ASE at:

National Institute for Automotive Service Excellence
101 Blue Seal Drive S.E.
Suite 101
Leesburg, VA 20175
Telephone: 703-669-6600
Fax: 703-669-6123
http://www.ase.com

2 | Take and Pass Every ASE Test

Participating in an Automotive Service Excellence (ASE) voluntary certification program gives you a chance to show your customers that you have the know-how needed to work on today's modern vehicles. The ASE certification tests allow you to judge your skills and knowledge against the automotive service industry's standards for each specialty area.

If you are the "average" automotive technician taking this test, you are in your mid-30s and have not attended school for about 15 years. That means you probably have not taken a test in many years. Some of you, on the other hand, may have attended college or taken post-secondary education courses and may be more familiar with taking tests and with test-taking strategies. There is, however, a difference between the educational tests you may be accustomed to and the ASE test you are preparing to take.

How Are the Tests Administered?

ASE administers its certification exams utilizing a Computer Based Testing (CBT) methodology. The CBT exams are administered at test centers across the nation.

While it is always recommended that you refer to the ASE website (www.ase.com) for the latest data regarding testing registration and exam dates, below is an overview of the available testing windows. CBT exams will be available four times annually, for two-month windows, with a month of no-testing in between each testing window.

- January/February—Winter CBT Testing Window

- April/May—Spring CBT Testing Window

- July/August—Summer CBT Testing Window

- October/November—Fall CBT Testing Window

Who Writes the Questions?

ASE test questions are written by service industry experts in the area being tested. Each area will have its own technical experts. Questions are entirely job related. They are designed to test the skills you need to be a successful technician. Theoretical knowledge is important and necessary to answer the questions, but ability to apply that knowledge is the basis of ASE test questions.

Each question has its roots in an ASE "item-writing" workshop, where service representatives from automobile manufacturers (domestic and import), aftermarket parts and equipment manufacturers, working technicians, and vocational educators meet to share ideas and translate them into test questions. Each test question written by these experts must survive review by all members of the group.

The questions are written to deal with the practical application of soft skills and knowledge of systems experienced by technicians in their day-to-day work. All questions are pretested and quality-checked by a national sample of technicians.

Those questions that meet ASE standards of quality and accuracy are included in the scored sections of the tests; the "rejects" are sent back to the drawing board or discarded altogether.

Each certification test is made up of between 40 and 80 multiple-choice questions.

Note, however, that each test could contain additional questions that are included for statistical research purposes only. Your answers to these questions will not affect your score, but since you do not know which ones they are, you should answer all questions in the test. The once-in-five-year recertification test will cover the same content areas as those listed in the preceding text. However, the number of questions in each content area of the recertification test will be reduced by about one-half.

Using multiple criteria, including cross-sections by age, race, and other background information, ASE is able to guarantee that a question does not bias for or against any particular group. A question that shows bias toward any particular group is discarded.

Objective Tests

A test is called an objective test if the same standards and conditions apply to everyone taking the test and there is only one correct answer to each question.

Objective tests primarily measure your ability to recall information. A well-designed objective test can also test your ability to understand, analyze, interpret, and apply your knowledge. Objective tests include true–false, multiple-choice, fill-in-the-blank, and matching questions. ASE's tests consist exclusively of four-part multiple-choice objective questions.

The following are some strategies that may be applied to taking your tests.

Before beginning an objective test, quickly look over the test to determine the number of questions, but do not try to read through all the questions. In an ASE test, there are usually between 40 and 80 questions, depending on the subject. Read through each question before marking your answer. Answer the questions in the order they appear on the test. Leave those questions blank that you are not sure of and move on to the next question. You can return to those unanswered questions after you have finished the others. They may be easier to answer at a later time after your mind has had additional time to consider them at a subconscious level. In addition, you might find information in other questions that will help you recall the answers to some of the unanswered ones.

Do not be obsessed by the apparent pattern of responses. For example, do not be influenced by a pattern like **D, C, B, A, D, C, B, A** on an ASE test.

There is also a lot of folk wisdom about taking objective tests. For example, there are those who would advise you to avoid response options that use words such as *all, none, always, never, must,* and *only,* to name a few. This, they claim, is because nothing in life is exclusive. They would advise you to choose response options that use words that allow for some exception, such as *sometimes, frequently, rarely, often, usually, seldom,* and *normally.* They would also advise you to avoid the first and last option (**A** or **D**) because test writers, they feel, are more comfortable if they put the correct answer in the middle (**B** or **C**) of the list of choices. Another recommendation often offered is to select the option that is either shorter or longer than the other three choices because it is more likely to be correct. Some would advise you to never change an answer since your first intuition is usually correct.

Although there may be a grain of truth in this folk wisdom, ASE test writers try to avoid them, and so should you. There are just as many **A** answers as there are **B** answers, and just as many **C** answers as **D** answers. As a matter of fact, ASE tries to balance the answers at about 25 percent per choice **A, B, C,** and **D**. There is no intention to use "tricky" words, such as outlined previously. Put no credence in the opposing words "sometimes" and "never," for example.

Multiple-choice tests are sometimes challenging because there are often several choices that may seem possible, and it may be difficult to decide on the correct choice.

The best strategy, in this case, is to first determine the correct answer before looking at the options. After arriving at the answer you have worked out, you should still examine the options given to make sure that none seems more correct than yours. If you do not know or are not sure of the answer, read each option carefully and try to eliminate those options that you know are incorrect. That way, you can often arrive at the correct choice through a process of elimination.

If you have gone through all the test questions and you still do not know the answer to some of the questions, then guess. Yes, guess. You then have at least a 25 percent chance of being correct. If you leave the question blank, you have no chance. Your score is based on the number of questions answered correctly.

Preparing for the Exam

The main reason we have included so many sample and practice questions in this guide is, simply, to help you learn what you know and what you don't know. We recommend that you work your way through each question in this book. Before doing this, carefully look through Section 3: It contains a description and explanation of the types of questions you'll find in an ASE exam.

Once you understand what the questions will look like, move to the sample test questions (Section 5). Read the explanations (Section 7) to the answer for each question, and if you don't feel you understand the reasoning for the correct answer, go back and read the overview (Section 4) for the task that is related to that question. If you still don't feel you have a solid understanding of the material, identify a good source of information on the topic, such as a textbook, and do some more studying.

After you have completed all the sample test items and reviewed your answers, move to the additional questions (Section 6). This time, answer the questions as if you were taking an actual test. Do not use any reference or allow any interruptions so as to get a feel for how you will do in an actual test. Once you have answered all the questions, grade your results using the answer keys in Section 7. For every question to which you gave an incorrect answer, study the explanations to the answers and/ or the overview of the related task areas (Section 4). Try to determine the root cause for your missing the question. The easiest thing to correct is learning the correct technical content. The hardest thing to correct is the behavior that leads you to the incorrect answer. If you knew the information but still got it incorrect, there is a behavior problem that will need to be corrected. An example would be reading too quickly and skipping over words, which affect your reasoning. If you can identify what you did that caused you to answer the question incorrectly, you can eliminate that cause and improve your score. Here are some basic guidelines to follow:

- Focus your studies on those areas you are weak in.
- Be honest with yourself while determining if you understand something.
- Study often but in short periods of time.
- Remove yourself from all distractions while studying.
- Keep in mind that the goal of studying is not just to pass the exam: The real goal is to learn!
- Prepare physically by getting a good night's rest before the test and have meals that provide energy but do not cause discomfort.
- Arrive early at the test site to avoid long waits as test candidates check in and to allow all the time available for your tests.

During the Test

When taking a CBT exam, as soon as you are seated in the testing center, you will be given a brief tutorial to acquaint you with the computer-delivered test, prior to taking your certification exam(s). The CBT exams allow you to only select one answer per question. You can also change your answers as many times as you like. When you select a second answer choice, the CBT will automatically unselect your first answer choice. If you want to skip a question to return to later, you can utilize their "flag" feature, which will allow you to quickly identify and review questions whenever you are ready. Prior to completing your exam, you will also be provided with an opportunity to review your answers and address any unanswered questions.

If you have finished answering all the questions on a test and have time remaining, go back and review the answers of those questions that you were not sure of. You can often catch careless errors by using the remaining time to review your answers. Carefully check your answer sheet for blank answer blocks or missing information.

At practically every test, some technicians will invariably finish ahead of time and turn their papers in long before the final call. Some technicians may be taking a recertification test and others may be taking fewer tests than you. Do not let them distract or intimidate you.

It is not wise to use less than the total time that you are allotted for a test. If there are any doubts, take the time for review. Any product can usually be made better with some additional effort. A test is no exception. It is not necessary to turn in your test paper until you are told to do so.

Your Test Results!

You can gain a better perspective about tests if you know and understand how they are scored. ASE's tests are scored by a non-partial, unbiased organization having no vested interest in ASE or in the automotive industry.

Each question carries the same weight as any other question. For example, if there are 50 questions, each is worth 2 percent of the total score. The passing grade is 70 percent. That means you must correctly answer 35 of the 50 questions to pass the test.

The test results can tell you (1) where your knowledge equals or exceeds that needed for competent performance or (2) where you might need more preparation.

Your ASE test score report is divided into content areas and will show the number of questions in each content area and how many of your answers were correct. These numbers provide information about your performance in each area of the test. However, because there may be a different number of questions in each content area of the test, a high percentage of correct answers in an area with few questions may not offset a low percentage in an area with many questions.

It should be noted that one does not "fail" an ASE test. The technician who does not pass is simply told "more preparation needed." Though large differences in percentages may indicate problem areas, it is important to consider how many questions were asked in each area. Since each test evaluates all phases of the work involved in a service specialty, you should be prepared in each area. A low score in one area could keep you from passing an entire test.

There is no such thing as average. You cannot determine your overall test score by adding the percentages given for each task area and dividing by the number of areas. It doesn't work that way because generally the number of questions in each task area is not the same. A task area with 20 questions, for example, counts more toward your total score than a task area with 10 questions.

Your test report should give you a good picture of your results and a better understanding of your strength and weaknesses for each task area.

If you fail to pass the test, you may take it again at any time it is scheduled to be administered. You are the only one who will receive your test score. Test scores will not be given over the telephone by ASE nor will they be released to anyone without your written permission.

3 Types of Questions on an ASE Exam

ASE certification tests are often thought of as being tricky. They may seem to be tricky if you do not completely understand what is being asked. The following examples will help you recognize certain types of ASE questions and avoid common errors.

Most initial certification tests are made up of 40–80 multiple-choice questions. Multiple-choice questions are an efficient way to test knowledge. To answer them correctly, you must think about each choice as a possibility, and then choose the one that best answers the question. To do this, read each word of the question carefully. Do not assume you know what the question is about until you have finished reading it.

About 10 percent of the questions on an actual ASE exam will use an illustration. These drawings contain the information needed to correctly answer the question. The illustration must be studied carefully before attempting to answer the question. Often, technicians look at the possible answers and then try matching the answers to the drawing. Always, however, do the opposite: match the drawing to the answers. When the illustration is showing an electrical schematic or another system in detail, look over the system and try to figure out how the system works before you look at the question and the possible answers.

Multiple-Choice Questions

The most common type of question used on ASE tests is the multiple-choice question. This type of question contains three "distracters" (incorrect answers) and one "key" (correct answer). When the questions are written, effort is made to make the distracters plausible to draw an inexperienced technician to one of them. This type of question gives a clear indication of the technician's knowledge. If you encounter a question that you are unsure of, reverse engineer it by eliminating the items that it cannot be. Consider the following example:

Which of the following would be the result of an electrical short to ground before the load?

A. Reduced current flow
B. Circuit protection device opens
C. Circuit operates normally
D. Voltage drop across the load is increased   (A7)

Answer A is incorrect. The current flow would be greatly increased, which would cause the circuit protection device to open.
Answer B is correct. A short to ground before the load would cause the electrical flow to run straight to ground and bypass the load in the circuit. This would cause the circuit resistance to drop very low, which would cause a dramatic increase in current flow, which would open the circuit protection device.
Answer C is incorrect. The circuit would not operate normally due to the shorted path to ground before the load.
Answer D is incorrect. The load in the circuit would not have any voltage applied to it due to the shorted path to ground before the load.

EXCEPT Questions

Another type of question used on ASE tests has answers that are all correct except one. The correct answer for this type of question is the answer that is incorrect. The word "EXCEPT" will always be in capital letters. You must identify which of the choices is the incorrect answer. If you read too quickly through the question, you may overlook what the question is asking and answer the question with the first correct statement. This will make your answer incorrect. An example of this type of question and the analysis is as follows:

All of the following electrical tools can be used as electrical conductors in a circuit EXCEPT:

A. Copper
B. Aluminum
C. Rubber
D. Gold   (A2)

Answer A is incorrect. Copper is an excellent conductor that is widely used to connect electrical circuits.
Answer B is incorrect. Aluminum is an excellent conductor that is often used to make electrical terminals.
Answer C is correct. Rubber is classified as an insulator, which resists electrical flow.
Answer D is incorrect. Gold is an excellent conductor and is sometimes used on the tips of critical electrical terminals to improve the quality of the connection.

Technician A, Technician B Questions

The type of question that is most popularly associated with an ASE test is the "Technician A says... Technician B says... Who is correct?" type. In this type of question, you must identify the correct statement or statements. To answer this type of question correctly, you must carefully read each technician's statement and judge it on its own merit to determine if the statement is true.

Sometimes, this type of question begins with a statement about some analysis or repair procedure. This is often referred to as the stem of the question and provides the setup or background information required to understand the conditions on which the question is based. This is followed by two statements about the cause of the concern, proper inspection, identification, or repair choices. You are asked whether the first statement, the second statement, both statements, or neither statement is correct. Analyzing this type of question is a little easier than the other types because there are only two ideas to consider, although there are still four choices for an answer.

Technician A, Technician B questions are really double true-or-false questions. The best way to analyze this type of question is to consider each technician's statement separately. Ask yourself, is A true or false? Is B true or false? Then select your answer from the four choices. An important point to remember is that an ASE Technician A, Technician B question will never have Technician A and B directly disagreeing with each other. That is why you must evaluate each statement independently.

An example of this type of question follows:

The headlights on a truck are very dim on high and low beams. Technician A says that a loose common ground connection for the headlights could cause this problem. Technician B says that a blown headlight fuse could cause this problem. Who is correct?

A. A only
B. B only
C. Both A and B
D. Neither A nor B   (E1)

Answer A is correct. A loose common ground connection for the headlights could cause this problem. The technician would notice that the voltage drop on the headlights would be less than system voltage. The problem could be isolated by checking the voltage drop in the power and ground side of the circuit.

Answer B is incorrect. A blown headlight fuse would cause the headlights to be totally inoperative.

Answer C is incorrect. Only Technician A is correct.

Answer D is incorrect. Technician A is correct.

Most Likely Questions

Most likely questions are somewhat difficult because only one choice is correct while the other three choices are nearly correct. An example of a most likely question is as follows:

An operator complains that the backup alarm sounds low and weak. Which of these is the most likely cause?

A. An open backup switch
B. A stuck-closed backup switch
C. A blown backup light bulb
D. A burnt contact on the backup alarm relay (B12)

Answer A is incorrect. An open backup switch would cause the alarm to be completely inoperative.

Answer B is incorrect. A stuck-closed backup switch would cause the alarm to sound continuously.

Answer C is incorrect. A blown backup light bulb would not likely cause any issues with the backup alarm.

Answer D is correct. The relay for the backup alarm delivers the required current needed to operate the alarm. If the "load side" contacts get burned or corroded, then current flow is decreased due to the increased electrical resistance in the circuit.

LEAST LIKELY Questions

Notice that in the most likely questions, there is no capitalization. This is not so with LEAST LIKELY-type questions. For this type of question, look for the choice that would be the LEAST LIKELY cause of the described situation. Read the entire question carefully before choosing your answer.

An example is as follows:

The fuse for the heater and AC blows every time the blower motor is turned on. Which of the following is the LEAST LIKELY cause?

A. A loose ground at the blower motor
B. A short to ground in the blower power circuit
C. A shorted blower motor
D. A locked up blower motor (F4)

Answer A is correct. A loose ground at the blower motor would cause the blower motor to run slower or not at all. It would not cause the fuse to blow because the circuit electrical resistance would be higher, which would cause current flow to decrease.

Answer B is incorrect. A short to ground before the load will cause a circuit protection device to open.

Answer C is incorrect. A shorted blower motor could cause the fuse to blow each time the blower is turned on.

Answer D is incorrect. A locked-up blower motor would cause the electrical current to spike and result in the fuse blowing.

Summary

There are no four-part multiple-choice ASE questions having "none of the above" or "all of the above" choices. ASE does not use other types of questions, such as fill-in-the-blank, completion, true–false, word-matching, or essay. ASE does not require you to draw diagrams or sketches. If a formula or chart is required to answer a question, it is provided for you. There are no ASE questions that require you to use a pocket calculator.

Testing Time Length

Each individual ASE CBT exam has a fixed time limit. Individual exam times will vary based upon exam area, and will range anywhere from a half hour to two hours. You will also be given an additional 30 minutes beyond what is allotted to complete your exams to ensure you have adequate time to perform all necessary check-in procedures, complete a brief CBT tutorial, and potentially complete a post-test survey. You should be on time to ensure that you have all of the allocated time available. If you arrive late for a CBT test appointment, you will only have the amount of time remaining in your appointment.

Visitors are not permitted at any time. If you wish to leave the testing room for any reason, you must first ask permission. Even if you finish your test early and wish to leave, you would be permitted to do so only during specified dismissal periods.

You should monitor your progress and set an arbitrary limit to how much time you will need for each question. This should be based on the number of questions you are attempting. It is suggested that you wear a watch because some facilities may not have a clock visible to all areas of the room.

4 Overview of the Task List

Auxiliary Power Systems Installation and Repair (Test E3)

The following section includes the task areas and task lists for this test and a written overview of the topics covered in the test.

The task list describes the actual work you should be able to do as a technician and on which you will be tested by the ASE. This is your key to the test, and you should review this section carefully. We have based our sample test and additional questions upon these tasks. The overview section will also support your understanding of the task list. ASE advises that the questions on the test may not equal the number of tasks listed; the task lists outline the specific tasks that ASE expects you to know how to perform and which you will be tested on.

At the end of each question in the Sample Test and Additional Test Questions sections, a letter and number will be used as a reference to this section for additional study. Note the following example:

A pressure-compensated hydraulic pump has inconsistent flow under load. Technician A claims that a restricted case drain hose could be the cause. Technician B claims that an undersized load sense hose could be the cause. Who is correct?

A. A only
B. B only
C. Both A and B
D. Neither A nor B

(A1. 3)

Answer A is incorrect. The case drain will not affect the pump operation.
Answer B is correct. The hose size will affect how quick the pump can react to loads and speeds required downstream.
Answer C is incorrect. Only Technician B is correct.
Answer D is incorrect. Only Technician B is correct.

Task List and Overview

A. Hydraulic Systems (30 Questions)

Task A1   Pumps (8 Questions)

Task A1.1   Determine pump types and rotation.

Properly identify pump types for correct installation techniques and verify the proper rotation of the input shaft. Hydraulic pumps used in mobile applications generate flow; how they generate the flow varies by the pump type. A gear pump or vane pump typically produces flow regardless of the power take off (PTO) speed, but a variable displacement pump light only generates flow when a load is realized at the pump. So the diagnosis procedure is affected. Also note, depending on the transmission application, installing a pump with the incorrect rotation can render the system inoperative.

Task A1.2   **Install pump properly to include spline lubrication, brackets/supports, location, driveshaft angles, slip joint location(s), case drain, and hydraulic connections.**

Properly install direct and remote mounted pumps to the transmission or the chassis. Identify the pump size and the appropriate brackets required for its size and weight. Correctly install the hydraulic hoses to the correct ports on the pump.

To prevent the possibility of speed fluctuations caused by shaft universal joints operating at an angle, proper phasing is required. Most PTO/pump shafts require phasing both shaft yokes on the same plane during installation. This ensures that the PTO and the pump will operate in unison, even though the shaft speed may fluctuate. Always check the manufacturer's service manuals to confirm the proper phasing procedure.

Task A1.3   **Diagnose causes of unusual pump noises, temperatures, and flow; determine needed repairs.**

Inspect the pumps for noises that can be attributed to cavitation, over-pressurization, and improper plumbing. Perform appropriate repairs as needed. When a hydraulic pump operates it performs two functions. It creates a vacuum to draw oil into the pump and generates oil flow out of the pump. Both of these activities require the rest of the system be designed correctly. Any loss of oil into the pump will damage the pump by creating friction/heat and increasing the wear/damage in the pump. Low oil-restricted suction filters and/or undersized hydraulic hoses may also cause the same effect on a pump. Pumps generate flow; they do not produce pressure. So pumps are directly affected by the system design and relief valve settings. Incorrectly adjusted relief valves may require the pump to operate beyond its pressure rating. This added pressure would create excessive wear in the pump seals and ultimately cause pump failure.

Task A1.4   **Verify proper fluid application.**

Verify the proper fluid viscosity in the system. Care must be taken to use the correct oil. Low fluid level will cause decreased performance and erratic operation. It may also cause a growling or cavitation noise within the pump. Foaming in the remote reservoir may indicate air (aeration) in the system. Most manufacturers recommend checking the hydraulic fluid level at operating or working temperatures of 100–160 degrees Fahrenheit.

Task A2   Filtration/Reservoirs (Tanks) (5 Questions)

Task A2.1   **Identify type of filtration system; verify filter application and flow direction.**

Review the correct filtration needs of the hydraulic system. Verify the correct size, flow, pressure, and filtering media. All hydraulic systems require clean oil to ensure the system operates properly. Any contamination of the oil will travel throughout the system and increase the chance for early system failure. Most hydraulic systems require a return filter, which captures contaminants that otherwise may enter the oil tank. They also have a suction strainer to protect the pump from any debris that may be in the oil tank. It is important that any hydraulic filter being replaced in a system meet the system's specifications. A return filter improperly sized, either by its GPM capacity or its filter element, can cause contaminats to pass through the filter and thus possibly damage system components. Also, contaminants a filter installed backwards may cause cavitations of the hydraulic pump or may create excessive pressure drop (also known as back pressure) and possibly overheat the system.

Task A2.2   **Install filter(s) in proper locations(s); flush system in accordance with manufacturers' recommendations.**

Install filters in the correct location(s); replace filters during normal PM services, or when contamination is found. Review proper methods to flush and drain hydraulic oil from the reservoirs and actuators. Proper location of the filters helps to readily identify the condition of the filter and ease in its maintenance. Replacing filters on a scheduled basis helps ensure the longevity of the system. If the hydraulic oil has been contaminated either by overheating or contamination of the system, the oil should be completely drained and flushed. If the contamination is serious enough, all the actuators may need to be flushed clean.

Task A2.3   **Diagnose cause(s) of system contamination.**

Review the hydraulic system for possible issues attributed to: aeration, cavitation, foaming, dirt, moisture, and overheating, which can be destructive to a hydraulic system. Hydraulic oil can be contaminated by both internal and external sources. Air ingested into a hydraulic pump may be caused by an undersized oil reservoir or not keeping the oil at the correct level. This can cause cavitations in the pump, of which the resulting wear from the pump travel through the rest of the system, thus damaging/clogging the other system components.

Task A2.4   **Service filters and breathers in accordance with manufacturers' recommendations.**

Perform periodical service on hydraulic system filtering components. Review duty cycles with operators for any changes. Air is required to ensure the oil flows properly into the pump. As an actuator uses the oil in the system, the level of the oil in the tank will drop. If the breather filter is clogged with dirt, it may create cavitation issues in the tank and starve the pump. As with any filters in the system, they should be properly reviewed and replaced as needed.

Task A2.5   **Install reservoirs/tanks, related components, and shut-off valves in accordance with recommended procedures; flush and clean as required.**

Install correctly sized reservoirs to prevent system turbulence and foaming. Install sight gauges and dip sticks to verify correct fluid level. Properly remove cleanout covers and use noncorrosive material to clean and flush out dirt and debris in tanks.

Task A3   Hoses, Fittings, and Connections (4 Questions)

Task A3.1   **Identify proper applications to include sizes, types, and pressure/flow ratings.**

Review and make sure the correct hose rating is used in the system. Ensure the proper SAE rating, pressure, and flow requirements are met. Improper hose selection can lead to excessive pressure drops and overheating issues within circuits. Also choosing the improper hose type can lead to collapsed hoses, creating cavitation in suction (inlet) circuits.

Task A3.2   **Determine hydraulic layout (length, size, routing, bend radii, and protection).**

When routing hoses and tubing, properly secure assemblies away from heat sources and moving components. This can reduce the life of the hose and possibly create leaks due to excessive movement. Install hoses without tight bends or twists in the lines. These can dramatically reduce the life expectancy by up to 90 percent.

Task A3.3   **Determine correct application of thread sealants.**

Review the four standard thread types and the pressure ratings for each. Review the correct sealant used for NPT threads. Most mobile hydraulic systems use one of four fittings types: NPT, JIC, SAE o-ring, and Flat Face fittings. Ensuring the correct fitting for the application is critical for the system to operate properly. Using an NPT pipe thread in a high-pressure application may create an oil leak under load. Also, since only the NPT fitting requires an external sealant, which may contaminate the hydraulic system if improperly applied, these are slowly becoming obsolete in the newer high-pressure systems. These are still in use in older low-pressure systems, so it is very important to use the correct sealant.

Task A3.4   **Assemble hoses, connectors, and fittings in accordance with manufacturers' specifications; use proper procedures to avoid contamination.**

Assemble and install hose assemblies without introducing any contamination that may take place during hose-fitting fabrication. Use compressed air to clean out assemblies and install plastic caps to reduce dirt and debris from entering the hoses prior to installation.

Task A4   Control Valves (2 Questions)

Task A4.1   Identify control valves (directional and accessory) application and porting.

Identify the control valve as a directional, flow, or pressure control. Verify the ports in the valve to ensure the correct SAE or NPT fittings are used to adapt to the hose. The directional valve is the key component in controlling the operation of the system. Proper identification of the valve will aid in diagnosing system problems. Visually it may be difficult to determine how a valve operates. Replacing a 4-way valve section with a 3-way section may cause the actuator to not operate properly. It is imperative to recognize any port relief that may be located in directional valves, as they act as limits to protect the pump actuator from over-pressurization. Also, always replace valves with correct porting.

Task A4.2   Install valves in accordance with recommended procedures regarding location, mounting, and shielding; verify flow direction.

Review the valve and its mounting location. Review the schematic and valve porting to ensure proper flow paths are maintained. Depending on the control mechanism of a valving, it may need to be located within easy access to the operator while at the same time be protected from road spray and debris. Another factor in valve mounting location is the plumbing to the valve. If the system requires large hose sizes, it may restrict areas where the valve can be located. Again, the system design will dictate the proper valve porting, and thus the plumbing, but we also need to be aware of how to connect the valve into the system.

Task A4.3   Verify system-operating pressures and flow; confirm component compatibility.

Review the system pressure and flow requirements before selecting new or replacement components. Verify that Closed Center and Open Center components are not mismatched. With the complex systems used on trucks mounted systems, it is important the technician refers to any available drawings or prints. Mismatching components can cause system damage. For example, if a Closed Center Valve is used on a standard gear pump system, it may cause the pump to seize as the pump flow will "dead head" by over-pressuring the pump and causing it to fail.

Task A4.4   Verify, install, and adjust valve controls (electrical, mechanical, and pneumatic).

Install and adjust the valves per system requirements. Properly lock down the adjustments to prevent any loosening or backing off of settings. When installing or adjusting a hydraulic component, it may require the technician to set pressure settings and any actuators. These setting must be done according to the system's design.

Task A5   Actuators (3 Questions)

Task A5.1   Purge/bleed system in accordance with recommended procedures.

Review the actuator bleeder screws' location and remove air from the component. Air in a cylinder may not be easily removed based on its location in the vehicle. The air in the actuator can create a spongy effect, as the air can be compressed while the oil cannot. In the case of a single-acting dump, it may require the technician to "bleed" the air out of the cylinder by loosening a hydraulic hose.

Task A5.2   Diagnose the cause of incorrect actuator movement; determine needed repair.

Visually review the actuator's movement and determine if the movement is too slow or fast due to system flow or incorrect settings. When replacing a directional valve, it can be difficult to readily identify the internal flow paths of the valve; so it is always important to test the actuator movement. For example, if a valve section was replaced, the technician needs to verify the cylinder moves in the proper direction; if not, the hose may be hooked up backwards or the mechanical linkage could be backwards.

Task A5.3   **Diagnose the cause of seal failure; determine needed repair.**

A damaged seal can cause internal and external leaks. Review the leaks that can cause actuator movement issues or system contamination. Seals are used throughout the hydraulic system and can be either internal or externally located. Not all seals perform the same function. Some are designed to hold in pressure/oil while others are designed to wipe/clean a rod end of a cylinder. If the cylinder rod has a knick or dent in the rod, it will wear and abrade the seal over time and allow contamination to enter the cylinder. It is always important to visually inspect these areas to prevent major repairs from occurring.

Task A5.4   **Identify hydraulic motor type and location.**

Review motor rotation for correct rotation and speed. Hydraulic motors are physically similar to pumps except that they convert oil flow to turn a mechanical device. The rotation of a motor is directly affected by the plumbing from the directional valve. The pressure relief settings must be checked to ensure the motor, or the mechanical device the motor is connected to, cannot be damaged.

Task A5.5   **Verify case drain operation (where applicable).**

When case drains are required, review proper hose size and installation procedures. Issues with improper case drains can cause seal damage and leaks. These drains are used on high flow devices, such as a hydraulic motor or a piston pump, where a buildup of oil inside the case may cause leakage. The case drain allows this extra oil to return to the tank with little or no resistance (back pressure). A hose sized too small may allow the pressure in the motor case to build up too high, and thus cause seal failure.

Task A5.6   **Identify cylinder type (single or double acting).**

Review the differences in single- and double-acting cylinders. Review the methods of plumbing and removing air in actuators. When a double-acting cylinder has been repaired, or replaced, it needs to be operated slowly upon initial start-up. This allows the technician to verify that the direction of movement corresponds with the label/decal associated with the directional valve, and it also allows any air trapped in the cylinders to be evacuated. On a single-acting cylinder, the cylinder may need to be manually bled by loosening the hose fitting.

Task A6   General System Operation (8 Questions)

Task A6.1   **Interpret system diagrams, schematics, and layouts.**

Review hydraulic symbols and their effect on the hydraulic circuit. Note the location of pressure relief valves in the diagram and the result of improper settings. On mobile hydraulic systems, it is very difficult to understand how a system is designed just by performing a visual review. Access to the hydraulic diagram and understanding their symbols will aid the technician's understanding of whether the slow actuator movement on a dump body is due to an improperly adjusted relief valve.

Task A6.2   **Identify proper tools for installation, diagnosis, maintenance, and repair.**

Use a Flow Meter and Pressure Gauge Set to properly diagnose system issues; an example of one issue would be loss of flow under load. Determine which issues can be attributed to internal component wear, such as damaged seals or worn housings. During a diagnostic check of a motor-driven circuit, the motor slows to a halt as the load increases. The use of a flow meter with a pressure gauge allows the technician to isolate the issue between an improperly adjusted relief valve and an excessive pressure drop on the return side of the motor.

Task A6.3   **Perform general service diagnosis and repair procedures.**

Perform visual inspections of the hydraulic system. Replace any items that indicate wear but have not yet failed completely. Visually inspecting actuators and directional valves may prevent a major system failure. For example, a rusty rod end of a cylinder may cause seal damage and if caught early enough may just require a minor repair rather than a complete rebuild.

Task A6.4   **Assemble and integrate system components.**

Always install components as per the manufacturers' recommendation, and if various components are required, always check for compatibility of components. When repairing or replacing a component on a hydraulic system, it is critical that it meets or exceeds what the designer of the system intended. If the item is not rated for the system flow or pressure-operating parameters, it may cause other issues in the long term.

B. Mechanical Systems (13 Questions)

Task B1   **Verify PTO type, location, and mounting clearance; remove cover and identify PTO drive gear location and type (spur or helical).**

Note the difference in mounting and installation techniques on: a Wire Shift, Air Shift, and Hot Shift PTOs. Note when spacers and filler blocks are required to install PTOs. Note LH or RH sides of the transmission installation procedures and determine possible ratio and/or rotational changes. Note the differences of LH or RH Spur and Helical cut gear compatibility in order to ensure correct installation/mesh of gears.

Task B2   **Install supplied gaskets, verify correct fasteners and lock-tabs; install PTO, torque to specifications.**

Review the gaskets and the effect they have upon the gear set (lash) when installed. Ensure that no more than three gaskets are used in order to prevent loss of proper backlash. Use filler blocks as required. In order to prevent contamination in the transmission, do not use a sealant on the Hot Shift PTO.

Task B3   **Remove, where applicable, PTO shifter cover and measure backlash with dial indicator; adjust as needed.**

Dial indicators are used on manual transmissions in order to verify that the PTO gear mesh has the proper backlash; settings should be between .08" to .12". Note on automatic transmissions, that the backlash is determined by the PTO manufacturers' design. The manufacturer's supplied gasket should not be changed or altered in any way.

Task B4   **Install lubrication line, if required.**

On Hot Shift PTOs, a lubrication line is required to lube the internal bearings, unlike a manual transmission, which typically uses the splash from the internal gears. These lube lines also might have restrictors or orifices used to restrict the amount of oil passing through them.

Task B5   **Install shift controls and fastener lock tabs, if required.**

Install the PTO shifter controls so they can be operated easily without excessive resistance. Bending the wire shift cable excessively or allowing too much slack in the cable or connector can cause the PTO not to shift or to jump out of gear. The fastener lock tabs help keep the PTO mounting fasteners from loosening during operation.

Task B6   **Refill transmission to proper lubrication level with transmission manufacturer's recommended lubricant.**

Depending on the transmission, oil might need to be added due to the addition of the PTO. A PTO mounted on the bottom side might require more oil than a LH "Hump-up" location. Always check that the fluid level(s) after installation is complete and always match the OEM specifications (angles of unit installation).

Task B7   **Test operation of PTO; check for unusual noises and leaks; check shifter operation.**

Always test the PTO in and out of gear, being sure to listen for any unusual noises. A whining sound can indicate a backlash set below specifications. A clattering sound can indicate a backlash set with excessive clearance. The shifter operation should be smooth and allow the shift fork to fully engage and disengage the gears.

Task B8   **Verify proper rotational direction of output shaft.**

The PTO output shaft rotation on a manual transmission is typically CCW and CW on an automatic transmission. The PTO output shaft is based on the rotation of the transmission drive gear inside the housing. When reviewing an application the technician must be aware of the proper rotation of the PTO and its effect on the hydraulic pump. For instance; a standard manual transmission would rotate the PTO in a counterclockwise rotation. If the technician replaces the pump, he will need to review the pump to ensure that it will match that rotation for correct operation. Failure to do so may result in damage to the pumps internal components.

Task B9   **Install all warning and instructional decals/labels.**

All operation and warning labels must be placed near the devices or component installed. This is in order for operators and technicians to become aware of fluid types, contaminations and/or operating danger or restrictions. For example, if a truck has a Caution Decal on the side, it indicates that there is a rotating driveline under the cab and the technician should not go under a vehicle with the hydraulics operating.

Task B10   **Determine PTO driveshaft length, application, and operating angles.**

In order to reduce wear on U-joints, drivelines should be kept a short as possible and the working angles should be as close to three degrees as possible. If the angle of the driveline is greater than three degrees, a review of the critical speed data from the driveline manufacturer is required to ensure proper operation.

Task B11   **Install PTO driveshaft; lubricate U-joints and slip joints; check that operating angles are within manufacturers' recommended specifications.**

Install the driveline with the proper operating angles; the input and output shafts need to be parallel with each other. Avoid zero-degree operating angles. The U-joints require a maximum three-degree working angle in order to ensure that the needle bearings rotate and do not suffer from premature wear.

Task B12   **Check for proper PTO shaft timing (phasing).**

The driveline needs to be installed with the U-joints in phase to prevent driveline vibration and premature failure. A properly balanced and phased driveline will extend the life of the driveline assembly.

Task B13   **Check belt tension and alignment of belt driven components.**

Underhood clutch pump pulleys need to be installed parallel to the OEM pulleys and adjusted to within manufacturers' guidelines in order to prevent loss of torque to the pump and to prevent belt damage (both may be caused by slippage and out-of-parallel condition). For example, if the underhood clutch belt is not in line with the drive pulley, it may cause premature belt wear as well as possibly bearing damage to the clutch pump. In severe cases, the belt may actually "jump" off the pulleys.

Task B14   **Diagnose the causes of vibrations in auxiliary power trains.**

Any vibration in the auxiliary power circuit needs to be diagnosed immediately. Vibrations or noises can be attributed to PTO, driveline angles, or even hydraulic pump cavitations. These can be transmitted through the entire mechanical driveline.

Task B15   Test operation of PTO speed controls.

In order to operate the hydraulic system at the proper speed, the engine speed "RPM" is programmed to ensure the PTO driveline rotates at the proper speed based on the system design. Incorrect settings can lead to transmission damage (automatic) or pump and hydraulic system damage if set too high.

Task B16   Install guarding, if required.

Guards are used to protect the operator from rotating items. They are also used to protect the hydraulic system components from heat and debris resulting when a vehicle is driven off road. The guards are required to be reinstalled whenever they are removed to access a component.

Task B17   Diagnose the cause of abnormal PTO/component wear.

The hydraulic system will transmit energy throughout the entire system. Even the mechanical device like the PTO can be affected. So shock loads, excessive pressures, or duty cycles can damage components within the system. When diagnosing a normal PTO gear wear, remember the cause can be something upstream in the system.

C. Pneumatic Systems (2 questions)

Task C1   Determine proper location for pressure protection valve(s); install valve(s), and check operation.

The air shift PTO uses air from the OEM brake storage system, so the Pressure Protection Valve needs to be installed directly at the tank to prevent any air loss to the brake system. This valve will also limit when the air is allowed to pass through to the PTO circuit. Typically this valve will not allow air past it till the OEM side reaches 70 psi. This pressure setting is used to disengage the trucks parking brakes.

Task C2   Determine proper location for the air activated control valve.

Air-operated control valves are designed to ease installation in the cab of a truck. These can be located near the operator for ergonomic operation. The drawback is the air lines and valves are susceptible to heat and damage and should be located as far away as possible from heat and road debris.

Task C3   Connect DOT specified lines, fittings, and hoses to air activated control valve(s); determine proper routing of air hoses.

Accessing or tapping into the OEM system is required; all these lines are required to be DOT rated or meet or exceed FMVSS standards. Always use the components from the kits the manufacturers supply, as they contain the approved items.

Task C4   Verify operation of pressure-reducing (air regulator) valve(s).

These devices are installed to protect the OEM air system from an air leak that may occur by not allowing air pressure to pass through them without reaching a preset threshold. The technician reviewing the proper operation of these air-controlled components must be aware of air leaks not only on the hydraulic system, but perhaps also on the truck side that supplies the air pressure.

5 Sample Test for Practice

Sample Test

Please note the letter and number in parentheses following each question. They match the task in Section 4 that discusses the relevant subject matter. You may want to refer to the overview using the cross-referencing key to help with questions posing problems for you.

1. A dump pump will not lift a loaded body, but operates normally while empty. Technician A says a clogged strainer could be the cause. Technician B says a relief valve needs adjustment. Who is correct?
   A. A only
   B. B only
   C. Both A and B
   D. Neither A nor B   (A1.4)

2. A PTO (power take off) mounted gear pump is making noise under load. What is the most likely cause?
   A. The suction line has a leak.
   B. The pressure line has a leak.
   C. The driveline is phased.
   D. The main relief is set too high.   (A1.3)

3. After replacing a hydraulic oil tank with a larger version to help manage the system temperature, Technician A says to label the tank to prevent accidental contamination. Technician B says the hydraulic system should be flushed and all the filters changed. Who is correct?
   A. A only
   B. B only
   C. Both A and B
   D. Neither A nor B   (B9. A2-5)

4. A gear pump is being replaced. Technician A says the pressure rating should be reviewed. Technician B says the rotation should be checked. Who is correct?
   A. A only
   B. B only
   C. Both A and B
   D. Neither A nor B   (A1.1)

5. A gear pump is being replaced and shows signs of pitting on the gears. Technician A recommends performing a vacuum check. Technician B recommends checking the tank strainer. Who is correct?
   A. A only
   B. B only
   C. Both A and B
   D. Neither A nor B   (A6.3)

6. While installing an underhood clutch pump hydraulic system, Technician A says the clutch should be burnished before operating the system. Technician B says the clutch pulley alignment should be checked with a straightedge. Who is correct?
   A. A only
   B. B only
   C. Both A and B
   D. Neither A nor B   (B13)

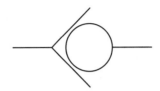

7. The symbol shown above is a:
   A. Hydraulic pump
   B. Pneumatic pump
   C. Check Valve
   D. Relief Valve   (A1.6)

8. When replacing a single-acting cylinder, Technician A says to remove the air from the cylinder. Technician B says to leak-check the rod seals. Who is correct?
   A. A only
   B. B only
   C. Both A and B
   D. Neither A nor B   (A5.1)

9. The air shift PTO (power take off) will not engage until the truck has run for 10 minutes. Technician A says a leaking air tank could be the cause. Technician B says a bad pressure protection valve could be the cause. Who is correct?
   A. A only
   B. B only
   C. Both A and B
   D. Neither A nor B   (C1)

10. A driver complains of jerky hydraulic operation. The mechanic notices the oil has a milky appearance to it. What is the most likely cause?
    A. Sticky control valve
    B. Leaking suction line
    C. Oil tank overfilled
    D. Dirty return filter   (A3.5)

11. A hydraulic gear pump has failed and needs to be replaced. All of these actions should be performed EXCEPT
    A. Replace return filter
    B. Flush transmission
    C. Check relief settings
    D. Change hydraulic oil   (A1.6)

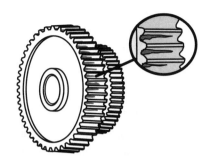

12. While disassembling a PTO (power take off), the damage to the PTO gear, as shown above, is seen. Technician A says too many PTO gaskets could be the cause. Technician B says excessive backlash could be the cause. Who is correct?
    A. A only
    B. B only
    C. Both A and B
    D. Neither A nor B   (B1)

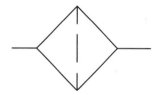

13. The symbol represented by the figure above is being replaced. All of these specifications should be checked to ensure that its replacement meets the system requirements EXCEPT
    A. Pressure ratings
    B. GPM (gallons per minute)
    C. SSU (Saybolt Seconds Universal)
    D. Micron rating   (A2.4)

14. A driver complains the PTO (power take off) powered lift-gate moves slower than it should during cold weather. What is the most likely cause?
    A. Leaking return filter
    B. Clogged breather
    C. Oversized return hose
    D. Low oil viscosity   (A2.4)

15. The hot shift PTO (power take off) engages slowly and slips under load. Technician A says the torque converter pressure could be too low. Technician B says a kinked shift hose could be the cause. Who is correct?
    A. A only
    B. B only
    C. Both A and B
    D. Neither A nor B   (B7)

16. Which of these fittings require a thread sealant?
    A. JIC (Joint Industry Council)
    B. NPT (National Pipe Thread)
    C. SAE (Society of Automotive Engineers)
    D. UNF (United Thread Standard)   (A3.3)

17. A hose assembly is being replaced. Technician A says the SAE (Society of Automotive Engineers) rating should match the application. Technician B says the diameter should match the application. Who is correct?
   A. A only
   B. B only
   C. Both A and B
   D. Neither A nor B   (A3.1)

18. A hydraulic pump will need to be remote mounted due to clearance issues. Technician A says to keep the driveline as short as possible to reduce vibration. Technician B says the pump mounting should be perpendicular to the frame. Who is correct?
   A. A only
   B. B only
   C. Both A and B
   D. Neither A nor B   (B10)

19. A hose is being replaced and the schematic indicates a -6 is required. What is this referring to?
   A. Burst pressure
   B. Inside diameter
   C. Outside diameter
   D. Working pressure   (A3.2)

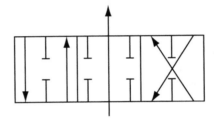

20. The symbol above represents a:
   A. 3-way. 3 position. Open center.
   B. 4-way. 3 position. Open center.
   C. 4-way. 2 position. Open center.
   D. 3-way. 2 position. Open center.   (A1.4)

21. A PTO (power take off) being installed has interference with the transmission housing and an idler spacer being installed. Technician A says the PTO shaft rotation has been altered. Technician B says the pump rotation will be affected. Who is correct?
   A. A only
   B. B only
   C. Both A and B
   D. Neither A nor B   (B1)

22. An operator complains of slow operation of the hydraulic auger. Which of these could be the cause?
   A. Oversized inlet line
   B. Excessive relief setting
   C. Excessive pressure drop
   D. Oversized case drain   (A6.3)

23. A hydraulic pump is being tested for loss of flow under load. Technician A says a flow meter should be plumbed in series. Technician B says a pressure should be plumbed into circuit. Who is correct?
   A. A only
   B. B only
   C. Both A and B
   D. Neither A nor B   (A6.3)

24. The internal bearing on a hot shift PTO (power take off) is showing signs of overheating and will be replaced. Technician A says the lubrication line could have an improper crimp. Technician B says the lube circuit could be missing the restrictor orifice. Who is correct?
   A. A only
   B. B only
   C. Both A and B
   D. Neither A nor B   (B4)

25. When installing a piston style pump, which of these actions should be done first?
   A. Bleed the load sense line
   B. Fill the housing with oil
   C. Set the low pressure standby
   D. Run the pump at 1.000 to seat the seals   (A2.2)

26. A 3-line pump is being converted from a 2-line configuration. Technician A says the extra line allows the body to rise faster. Technician B says the sleeve changes the flow inside the pump cavity. Who is correct?
   A. A only
   B. B only
   C. Both A and B
   D. Neither A nor B   (A1.1)

27. The air shifted PTO (power take off) will not properly shift in cold weather. Technician A says a water separator could be added to the air system. Technician B says the PTO air line should be relocated to another tank supply. Who is correct?
   A. A only
   B. B only
   C. Both A and B
   D. Neither A nor B   (C2)

28. The pressure of a hydraulic system is a result of:
   A. Main relief valve
   B. Pump displacement
   C. Actuator load
   D. Cylinder displacement   (A6.3)

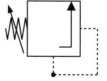

29. The symbol above is a:
   A. 4-way section
   B. Relief valve
   C. Check valve
   D. Flow divider   (A6.1)

30. After performing a PTO (power take off) installation the dial indicator shows 0.012". The technician should:
    A. Add 0.010 gasket
    B. Remove a 0.010 gasket
    C. Bend the lock tabs on nuts
    D. Fill the transmission with ATF (automatic transmission fluid)   (B2)

31. A hydraulic oil leak has developed under the truck chassis. The exact location is difficult to locate. In order to find the leak, Technician A recommends adding a non-corrosive dye to the oil. Technician B recommends running your hands along the hose. Who is correct?
    A. A only
    B. B only
    C. Both A and B
    D. Neither A nor B   (A6.3)

32. A closed center pump has a leak at the inlet seal. What is the most likely cause?
    A. A clogged case drain
    B. A clogged return filter
    C. A restricted inlet line
    D. A restricted pressure hose   (A5.5)

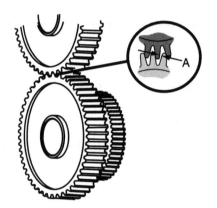

33. The preferred manner to check the PTO (power take off) gear set above, as indicated by "A," is with a:
    A. Feeler gauge
    B. Dial indicator
    C. Slide calipers
    D. Depth gauge   (B3)

34. The micron rating of a hydraulic return filter refers to its:
    A. Flow capability
    B. Beta measurement
    C. Filtering capability
    D. Absorption measurement   (A2.2)

35. A hydraulic system is being flushed due to contamination. Technician A recommends replacing all the filters. Technician B recommends filtering the new oil. Who is correct?
    A. A only
    B. B only
    C. Both A and B
    D. Neither A nor B   (A2.2)

36. A PTO (power take off) driveline vibration is felt when the circuit is running at the stepped up RPM (revolutions per minute). Technician A says the use of a solid shaft could be the cause. Technician B says the slip joint could be the cause. Who is correct?
    A. A only
    B. B only
    C. Both A and B
    D. Neither A nor B   (B14)

37. The hot shift PTO (power take off) pump has been replaced on an automatic transmission before starting the hydraulic system. Technician A recommends checking PTO backlash with a dial indicator. Technician B recommends backing off the relief valves. Who is correct?
    A. A only
    B. B only
    C. Both A and B
    D. Neither A nor B   (A4.4)

38. A customer complains that a newly installed hydraulic body dump will not lift enough. Technician A recommends increasing the main relief PSI (pounds per square inch) setting. Technician B recommends changing to a higher rated pump. Who is correct?
    A. A only
    B. B only
    C. Both A and B
    D. Neither A nor B   (A6.4)

39. A hot shift PTO (power take off) will not shift into gear after installation. Technician A says the shifter hose could be tapped into the wrong port. Technician B says the shift solenoid could be plumbed incorrectly. Who is correct?
    A. A only
    B. B only
    C. Both A and B
    D. Neither A nor B   (B7)

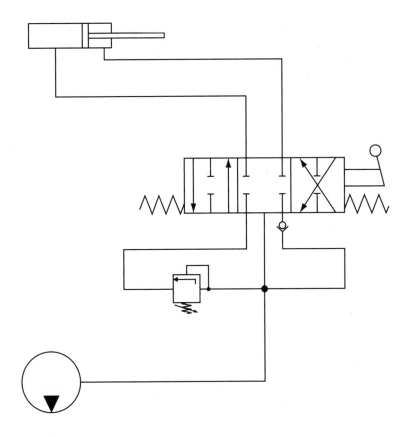

40. The circuit above drifts down under a load. What is the most likely cause?
    A. Leak on the rod end of the cylinder
    B. The main relief is set too low
    C. Sticking directional valve
    D. Worn pump gears   (A6.3)

41. After replacing a leaking hose, the operator complains of oil leaking from the top of the tank.
    What is the most likely cause?
    A. Debris in the main relief
    B. Clogged inlet filter
    C. Over-filled reservoir
    D. Over-speeding pump   (A2.3)

42. The operator of the PTO (power take off) driven blower system complains of vibration during
    operation. Technician A says the engine RPM (revolutions per minute) could be the issue.
    Technician B says the driveline working angles could be the issue. Who is correct?
    A. A only
    B. B only
    C. Both A and B
    D. Neither A nor B   (B12)

43. The oil tank breather is designed for all these issues EXCEPT
    A. Prevent oil leakages
    B. Maintain pump prime
    C. Reduce moisture
    D. Reduce tank turbulence   (A2.1)

44. An articulating crane is making significant amounts of noise while operating under high-speed conditions. What is the most likely cause?
 A. Overfilled tank
 B. Clogged return filters
 C. Incorrect hose fittings
 D. Excessive relief setting
(A3.1)

45. When programming the engine speed parameters for a PTO (power take off) set speed, Technician A says the engine should be programmed for torque converter lock up. Technician B says the PTO ratio should be checked. Who is correct?
 A. A only
 B. B only
 C. Both A and B
 D. Neither A nor B
(B15)

46. The integral main relief valve of a manifold block has been replaced. Before operating the system, the technician should:
 A. Set engine to Set PTO (power take off) RPM (Revolutions per minute)
 B. Turn down the adjustment
 C. Connect vent line to tank
 D. Bleed air from pressure line
(A4.4)

47. The hydraulic system was converted over to biodegradable hydraulic oil. Technician A says the relief valves should all be readjusted. Technician B says one of the benefits is that oil leakages do not require clean up. Who is correct?
 A. A only
 B. B only
 C. Both A and B
 D. Neither A nor B
(A1.4)

48. Technician A says a tandem pump can have two outlet ports. Technician B says a tandem pump can share an inlet port. Who is correct?
 A. A only
 B. B only
 C. Both A and B
 D. Neither A nor B
(A1.1)

49. While testing the operation of a pump on a single-acting dump truck, the engine stalls while raising the body. Technician A says the cylinder rod work port relief could be the cause. Technician B says excessive pump flow could be the cause. Who is correct?
 A. A only
 B. B only
 C. Both A and B
 D. Neither A nor B
(A6.3)

50. A winch motor has a kinked case drain hose. Technician A says to check the shaft seal for damage. Technician B says to flush the motor housing of contaminants. Who is correct?
 A. A only
 B. B only
 C. Both A and B
 D. Neither A nor B
(A5.5)

6 Additional Test Questions for Practice

Additional Test Questions

Please note the letter and number in parentheses following each question. They match the task in Section 4 that discusses the relevant subject matter. You may want to refer to the overview using the cross-referencing key to help with questions posing problems for you.

1. The driver complains that the rotation speed of a boom crane is too slow. Technician A says a kinked hose could be the cause. Technician B says a main relief could be the cause. Who is correct?
   A. A only
   B. B only
   C. Both A and B
   D. Neither A nor B
   (A5.2)

2. A pump's GPM (gallons per minute) rating is used to determine all of these EXCEPT:
   A. Load
   B. Speed
   C. Hoses
   D. Valves
   (A1.1)

3. A dump truck with a double-acting cylinder has developed an oil leak at the rod end of the cylinder. Technician A says to check the o-ring at the head end. Technician B says to check the cylinder bleeder screw. Who is correct?
   A. A only
   B. B only
   C. Both A and B
   D. Neither A nor B
   (A5.3)

4. A hydraulic manifold valve and hoses are being replaced. Technician A says incorrect fitting selection could cause restriction. Technician B says a hose twist could cause premature failure. Who is correct?
   A. A only
   B. B only
   C. Both A and B
   D. Neither A nor B
   (A3.2)

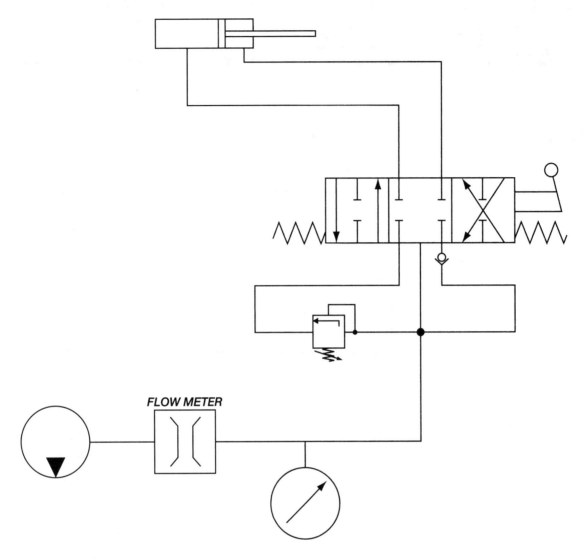

5. While testing the circuit above, the technician notices the flow drops when the pressure is raised on the system. What is the most likely cause?
   A. Inline relief
   B. Pump seal
   C. Cylinder seal
   D. Valve sticking   (A6.3)

6. A double-acting cylinder lowers erratically when the body is unloaded. Technician A says a leaking main relief could be the cause. Technician B says a valve spool could be the cause. Who is correct?
   A. A only
   B. B only
   C. Both A and B
   D. Neither A nor B   (A5.3)

7. When replacing a damaged hose assembly, Technician A says a hose twist can cause the hose to burst. Technician B says the hose should exceed the main relief. Who is correct?
   A. A only
   B. B only
   C. Both A and B
   D. Neither A nor B   (A3.4)

8. A pressure hose failed near the catalytic converter. Technician A says a missing heat shield might be the cause. Technician B says a tight hose bend radius might be the cause. Who is correct?
   A. A only
   B. B only
   C. Both A and B
   D. Neither A nor B   (A3.5)

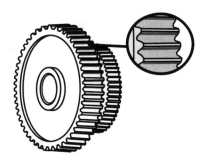

9. The PTO (power take off) drive gear is exhibiting the wear pattern shown above. What is the most likely cause?
   A. Incorrect lubricant
   B. Excessive backlash
   C. Over-torqued housing
   D. Unbalanced driveline   (B3)

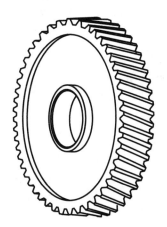

10. The PTO (power take off) gear shown above is a _____ design.
    A. RH (Right Hand)
    B. LH (Left Hand)
    C. Spur
    D. Sprocket   (B1)

11. The driver complains of a whining noise while operating the PTO (power take off). What is the most likely cause?
    A. Broken PTO gear
    B. Too much backlash
    C. Too little backlash
    D. Broken shift fork   (B3)

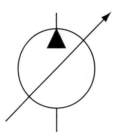

12. Technician A says the figure above indicates GPM (gallons per minute) is adjustable.
    Technician B says the figure above indicates PSI (pounds per square inch) is adjustable. Who is
    correct?
    A. A only
    B. B only
    C. Both A and B
    D. Neither A nor B   (A1.1)

13. The technician is checking for a leaking PTO (power take off) output shaft. What is the most
    likely cause?
    A. Excessive backlash
    B. Broken driven gear
    C. Unbalanced drive shaft
    D. Worn PTO shifter   (B17)

14. A PTO (power take off) driveline is exhibiting premature wear of the U-joints. Technician
    A says the operating RPM (revolutions per minute) should be checked. Technician B says the
    driveline balance should be checked. Who is correct?
    A. A only
    B. B only
    C. Both A and B
    D. Neither A nor B   (E4)

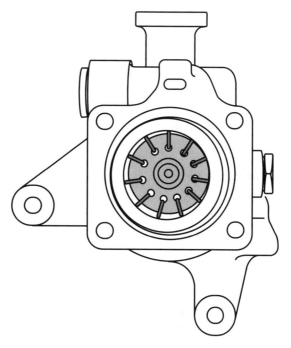

15. The pump above is a:
    A. Gear
    B. Vane
    C. Lobe
    D. Clutch   (A1.1)

16. The hydraulic system is overheating and causing poor performance. Technician A says an
    improperly set relief valve could be the cause. Technician B says incorrect fluid viscosity could
    be the cause. Who is correct?
    A. A only
    B. B only
    C. Both A and B
    D. Neither A nor B   (A6.3)

17. A PTO (power take off) pump assembly is leaking from the automatic transmission housing.
    Technician A says the pump might need to be remote mounted. Technician B says the PTO
    gasket might need additional sealant. Who is correct?
    A. A only
    B. B only
    C. Both A and B
    D. Neither A nor B   (A1.2)

18. A NPT (national pipe thread) fitting is being replaced in the electric-operated manifold
    valve. Technician A says liquid TFE (liquid Teflon sealant) should be applied to the threads.
    Technician B says Teflon tape should be wrapped 1/8" back on threads. Who is correct?
    A. A only
    B. B only
    C. Both A and B
    D. Neither A nor B   (A3.3)

19. The symbol above is a:
    A. Flow divider
    B. Oil cooler
    C. Return filter
    D. Motor   (A6.1)

20. A pressure-compensated hydraulic pump has inconsistent flow under load. Technician A says a restricted case drain hose could be the cause. Technician B says an undersized load sense hose could be the cause. Who is correct?
    A. A only
    B. B only
    C. Both A and B
    D. Neither A nor B   (A1.3)

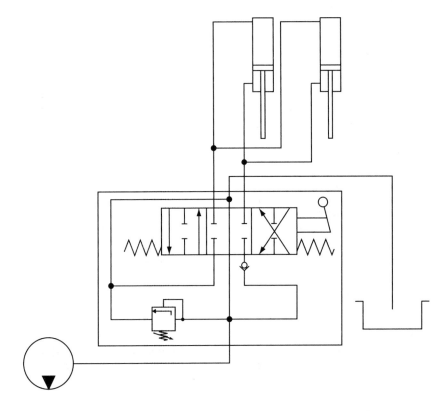

21. The pump shaft has been broken in the circuit above. What is the most likely cause?
    A. Bent cylinder rod
    B. Bad port relief
    C. Stuck valve spool
    D. Clogged return filter   (A6.3)

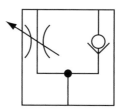

22. Technician A says the valve shown above can be used to regulate actuator speed. Technician B says the valve shown can be used to set actuator pressure. Who is correct?
    A. A only
    B. B only
    C. Both A and B
    D. Neither A nor B   (A6.1)

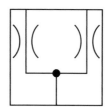

23. The component symbol shown above is a:
    A. Counterbalance valve
    B. 3-way port relief
    C. Flow divider
    D. Shuttle valve   (A6.1)

24. A dump pump was replaced in the field; the driver still complains it will not lift to capacity. Technician A says the new pump's displacement could be the issue. Technician B says the cylinder port relief valve could be the issue. Who is correct?
    A. A only
    B. B only
    C. Both A and B
    D. Neither A nor B   (A4.3)

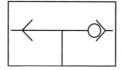

25. The symbol above is a:
    A. Dual check valve
    B. Flow divider
    C. Shuttle valve
    D. Counterbalance valve   (A6.1)

26. The driver complains the plow cylinder moves while the shifter is in neutral position. Technician A says the main relief could be stuck open. Technician B says a leaking cylinder spool could be the cause. Who is correct?
    A. A only
    B. B only
    C. Both A and B
    D. Neither A nor B   (A5.2)

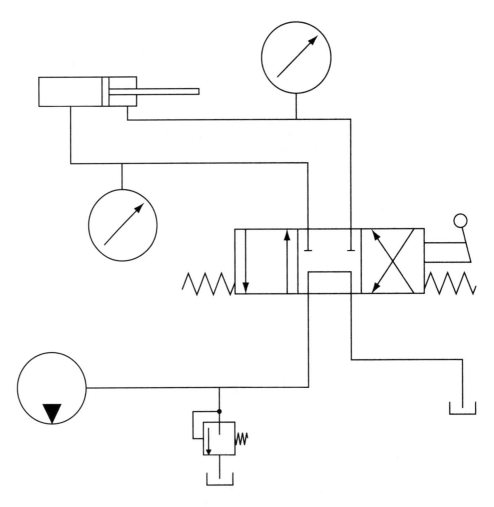

27. While checking the system for actuator operation, each pressure gauge reads 500PSI, as shown above. Technician A says this could cause the rod end of the cylinder to retract. Technician B says the gland packing could have an internal leak. Who is correct?
    A. A only
    B. B only
    C. Both A and B
    D. Neither A nor B   (A5.2)

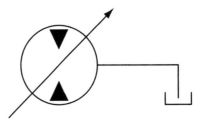

28. Technician A says the arrow indicates the operation can be reversed in the item above. Technician B says the pump speed can be changed/adjusted. Who is correct?
    A. A only
    B. B only
    C. Both A and B
    D. Neither A nor B   (A6.1)

29. A direct mount gear pump is being installed. Technician A says the pump splines should be lubricated. Technician B says the pump should weigh less than 50 lbs (pounds). Who is correct?
    A. A only
    B. B only
    C. Both A and B
    D. Neither A nor B   (A1.2)

30. After installing a new pump on a truck, none of the hydraulics will operate. Technician A says the pump's rotation could be incorrect. Technician B says the pump hoses could be installed backwards. Who is correct?
    A. A only
    B. B only
    C. Both A and B
    D. Neither A nor B   (A1.1, A1.2)

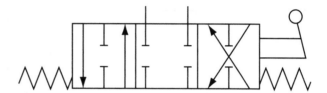

31. Technician A says the figure shown above indicates a 4-way, 3-section valve. Technician B says the figure indicates the valve has a motor spool. Who is correct?
    A. A only
    B. B only
    C. Both A and B
    D. Neither A nor B   (A4.1)

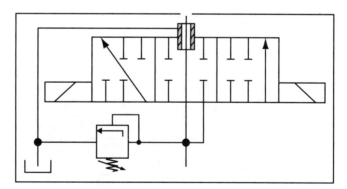

32. The 4-way valve above has all these features EXCEPT:
    A. Mechanical shift
    B. Internal relief
    C. Power beyond port
    D. Return port   (A4.1)

33. A single-acting cylinder was repacked and installed and now there is air trapped inside it. Technician A says to cycle the body up and down to remove the air pocket. Technician B says to open the bleeder screw on the rod end of the cylinder. Who is correct?
    A. A only
    B. B only
    C. Both A and B
    D. Neither A nor B   (A5.1)

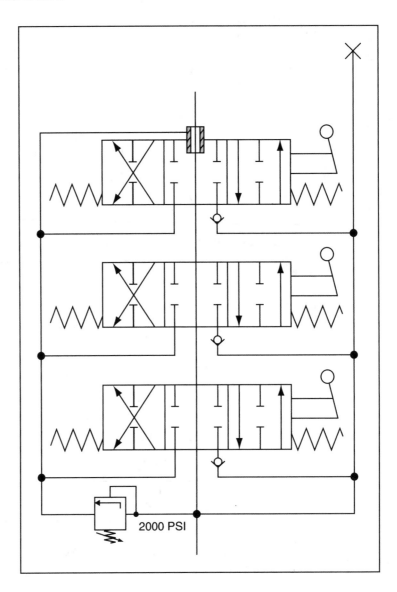

34. Technician A says the internal relief in the valve above can be used to protect the pump. Technician B says this valve in the figure above can be used to operate double-acting cylinders. Who is correct?
    A. A only
    B. B only
    C. Both A and B
    D. Neither A nor B   (A4.1)

35. A tank inlet hose is being replaced. Which of these is the correct hose?
    A. 100R1
    B. 100R2
    C. 100R3
    D. 100R4   (A3.1)

36. When installing a hot shift PTO (power take off), Technician A says the supplied gasket is used to set the proper backlash. Technician B says the silicon sealant is applied to both the gasket and the housing. Who is correct?
    A. A only
    B. B only
    C. Both A and B
    D. Neither A nor B   (B2)

37. The hydraulic system is operating at a high temperature. Technician A says the main relief spring could be the cause. Technician B says the oil could break down and damage the pump. Who is correct?
    A. A only
    B. B only
    C. Both A and B
    D. Neither A nor B   (A4.3)

38. The hydraulic winch motor is operating slowly and is not lifting to its capacity. Technician A says the motor housing could be worn. Technician B says the relief valve seat could be scored. Who is correct?
    A. A only
    B. B only
    C. Both A and B
    D. Neither A nor B   (A5.5)

39. The hydraulic system is being flushed and the fluid replaced due to excessive heat damage. Technician A says the hydraulic filters should be changed. Technician B says the relief valves might need to be replaced. Who is correct?
    A. A only
    B. B only
    C. Both A and B
    D. Neither A nor B   (A6.3)

40. The main return line to the tank has burst open. What is the most likely cause?
    A. Leaking return filter
    B. Clogged tank breather
    C. Worn filter bypass spring
    D. Oil viscosity too thick   (A3.4)

41. When assembling and installing the pressure hose on a 3-line dump truck, Technician A says to blow out the hose assembly before installing. Technician B says to protect the hose assembly from sharp edges. Who is correct?
    A. A only
    B. B only
    C. Both A and B
    D. Neither A nor B   (A3.4)

42. A disassembled gear pump has deep gouges in the gear paths. Technician A says the system could be experiencing shock loads. Technician B says it could be the return filters are bypassing. Who is correct?
    A. A only
    B. B only
    C. Both A and B
    D. Neither A nor B   (A1.3)

43. The PTO (power take off) driveline has a pronounced vibration when operating at PTO speeds. Technician A says this could cause a PTO leak. Technician B says this could cause an input shaft leak. Who is correct?
   A. A only
   B. B only
   C. Both A and B
   D. Neither A nor B   (B14)

44. A dump truck is being readied for seasonal service after long-term storage. Technician A says the hydraulic system should be started with no load. Technician B says all the actuators should be cycled thoroughly. Who is correct?
   A. A only
   B. B only
   C. Both A and B
   D. Neither A nor B   (A6.3)

45. When removing the pump from a hot shift PTO (power take off), the technician notices worn splines on the pump shaft. Technician A says contaminated oil could be the cause. Technician B says excessive load could be the cause. Who is correct?
   A. A only
   B. B only
   C. Both A and B
   D. Neither A nor B   (A1.2)

46. A double-acting dump cylinder will not retract. Technician A says a worn port relief spring could be the cause. Technician B says a bent/damaged cylinder rod could be the cause. Who is correct?
   A. A only
   B. B only
   C. Both A and B
   D. Neither A nor B   (A5.2, A6.3)

47. A directional valve has developed a leak in between the sections. Technician A says a blocked power beyond port could be the cause. Technician B says a relief valve stuck open could be the cause. Who is correct?
   A. A only
   B. B only
   C. Both A and B
   D. Neither A nor B   (A4.1)

48. When adjusting the relief valves in the hydraulic system, Technician A says the main relief should be set lower than the port relief. Technician B says the relief valves should be adjusted at operating temperature. Who is correct?
   A. A only
   B. B only
   C. Both A and B
   D. Neither A nor B   (A4.3)

49. The underhood pump is making noise upon its daily start-up. Technician A says the suction hose could be too short. Technician B says the oil tank could be raised. Who is correct?
   A. A only
   B. B only
   C. Both A and B
   D. Neither A nor B   (A2.5)

50. There is foaming occurring at the hydraulic reservoir sight glass. What is the most likely cause?
    A. Oversized reservoir
    B. Low hydraulic oil
    C. Restricted inlet screen
    D. Excessive pressure setting   (A2.3)

51. The auger motor on a spreader circuit is very hot to the touch after operating for only a half hour. Technician A says too many street elbows could be the cause. Technician B says the hose diameter could be too small. Who is correct?
    A. A only
    B. B only
    C. Both A and B
    D. Neither A nor B   (A4.3)

52. A properly designed oil tank should have the following EXCEPT?
    A. Fill screen
    B. Magnet plug
    C. Clean out cover
    D. Desiccant filter   (A2.5)

53. When replacing hose assemblies on a truck that have been damaged and to prevent future issues, Technician A says to route pressure and return lines separately. Technician B says to keep the hoses away from sharp edges. Who is correct?
    A. A only
    B. B only
    C. Both A and B
    D. Neither A nor B   (A3.2)

54. A new single-acting cylinder has been installed, but while testing the empty body it will not fully extend. Technician A says the main relief could be the cause. Technician B says the oil level could be too low. Who is correct?
    A. A only
    B. B only
    C. Both A and B
    D. Neither A nor B   (A6.3)

55. The oil in the tank smells burnt on a hydraulic asphalt dump body. Technician A says the oil viscosity could be the cause. Technician B says the duty cycle of the truck should be changed. Who is correct?
    A. A only
    B. B only
    C. Both A and B
    D. Neither A nor B   (A6.3, A6.4)

56. A PTO (power take off) driveline is being installed on a dump truck. Technician A says warning labels should be installed on both sides of the truck. Technician B says the driveline should not be running while servicing the unit. Who is correct?
    A. A only
    B. B only
    C. Both A and B
    D. Neither A nor B   (B9)

57. The slip joint is a part on the PTO (power take off) driveline that:
    A. Reduces vibration to the transmission
    B. Should be mounted on the pump end
    C. Supports the center bearing
    D. Allows the length to change   (B11)

58. When installing a bottom mount PTO (power take off) to the transmission, Technician A says no more than 3 gaskets should be installed. Technician B says the backlash should be set to 0.012" to 0.018". Who is correct?
    A. A only
    B. B only
    C. Both A and B
    D. Neither A nor B   (B2)

59. The hydraulic system of a closed loop system is running hot after a PTO (power take off) gear was replaced. What is the most likely cause?
    A. Improper PTO backlash
    B. PTO ratio too low
    C. PTO ratio too high
    D. Improper PTO oil viscosity   (B7)

60. Technician A says the item above is measured in BTU (British Thermal Units). Technician B says the item above is measured in microns. Who is correct?
    A. A only
    B. B only
    C. Both A and B
    D. Neither A nor B   (A6.1)

61. A bottom mount PTO (power take off) has a cracked housing. Technician A says excessive backlash could be the cause. Technician B says the lock tabs could be loose. Who is correct?
    A. A only
    B. B only
    C. Both A and B
    D. Neither A nor B   (B2)

62. There is excessive clutch plate wear in a hot shift PTO (power take off) that is causing it to slip under load. Technician A says torque converter pressure could be the cause. Technician B says lube line pressure could be the cause. Who is correct?
    A. A only
    B. B only
    C. Both A and B
    D. Neither A nor B   (B17)

63. The PTO (power take off) pump on an automatic transmission will not fully raise a load. Technician A says an obstructed PTO shift hose could be the cause. Technician B says the incorrect PTO set speed could be the cause. Who is correct?
    A. A only
    B. B only
    C. Both A and B
    D. Neither A nor B   (B5, B15)

64. The oil is foaming from the transmission dipstick equipped with a hot shift PTO (power take off). Technician A says the transmission cooler could be the cause. Technician B says the incorrect PTO ratio could be the cause. Who is correct?
    A. A only
    B. B only
    C. Both A and B
    D. Neither A nor B   (B6)

65. A cable shift PTO (power take off) slips out of gear while operating the dump body. Technician A says a loose shifter cover could be the cause. Technician B says the clutch adjustment could be the cause. Who is correct?
    A. A only
    B. B only
    C. Both A and B
    D. Neither A nor B   (B7)

66. The technician is installing hot shift PTO (power take off) on an automatic transmission that has multiple mounting locations. Before installation, all of these items should be considered EXCEPT:
    A. Exhaust system
    B. Pump rotation
    C. Frame clearance
    D. Gear ratio   (B1)

67. While running a hydraulic system under load, the PTO (power take off) driveshaft fell off the PTO shaft. Technician A says the slip joint clearance was insufficient and could be the cause. Technician B says if set screws were not locked in with safety wire it could be the cause. Who is correct?
    A. A only
    B. B only
    C. Both A and B
    D. Neither A nor B   (B10)

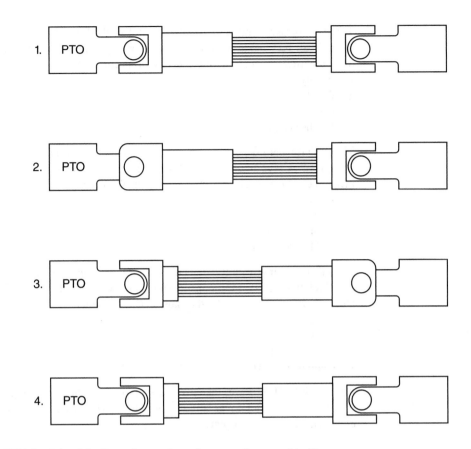

68. Which of the drivelines shown above is correctly assembled?
    A. 1
    B. 2
    C. 3
    D. 4
(B12)

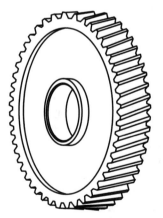

69. The gear shown above is located in the transmission housing. What type of PTO (power take off) gear should be installed?
    A. LH (Left Hand) helical
    B. Hypoid
    C. RH (Right Hand) helical
    D. Spur
(B3)

70. The PTO (power take off) is making a clattering sound when engaged but goes away when the clutch pedal is depressed. Technician A says excessive pump shaft resistance could be the cause. Technician B says the PTO gear backlash is 0.008" and could be the cause. Who is correct?
   A. A only
   B. B only
   C. Both A and B
   D. Neither A nor B   (B3)

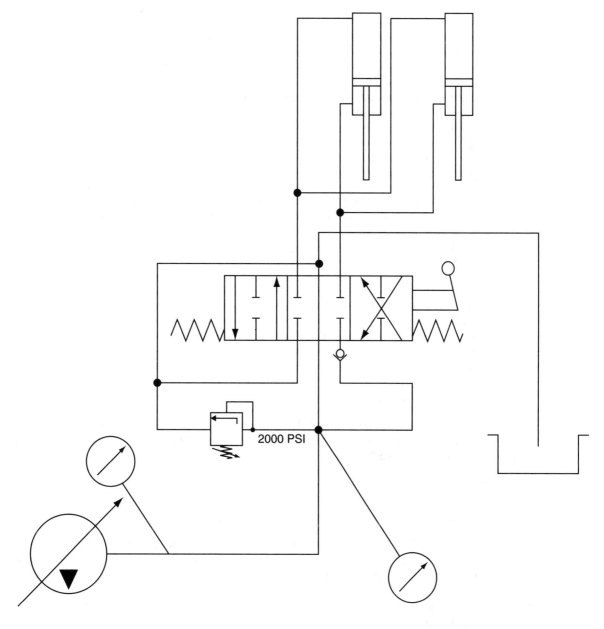

71. Pressure gauges are tapped in various points in the circuit above. What is the technician checking?
   A. Pump flow
   B. Main relief setting
   C. Pressure drop
   D. Low pressure standby   (A6.2)

72. The underhood pump bearing has failed. What is the most likely cause?
    A. Belt is misaligned
    B. Belt is too tight
    C. Worn belt tensioner
    D. Worn pulley groove   (B12)

73. When installing a hot shift PTO (power take off) on an automatic transmission, Technician A says a lube line is not needed when the PTO is mounted below the fluid line. Technician B says the lube line should be rated to the system pressure. Who is correct?
    A. A only
    B. B only
    C. Both A and B
    D. Neither A nor B   (B4)

74. Technician A says improper u-joint angles can damage seals. Technician B says improper grease can damage the U-joints. Who is correct?
    A. A only
    B. B only
    C. Both A and B
    D. Neither A nor B   (B11)

75. The technician is checking the underhood pump belts for tension. Technician A says a loose belt could jump off pulley. Technician B says a loose belt could slip under load. Who is correct?
    A. A only
    B. B only
    C. Both A and B
    D. Neither A nor B   (B13)

76. The operator of the truck complains of a vibration in the drivetrain when operating the hydraulic system. Technician A says the U-joints could be out of phase. Technician B says the driveline could be out of balance. Who is correct?
    A. A only
    B. B only
    C. Both A and B
    D. Neither A nor B   (B12, B14)

77. The PTO (power take off) set speed is being programmed. Technician A says the proper setting allows torque converter lockup. Technician B says the proper setting creates the correct pump flow. Who is correct?
    A. A only
    B. B only
    C. Both A and B
    D. Neither A nor B   (B15)

78. When installing a PTO (power take off) with a direct mounted pump, Technician A says the appropriate grease should be applied to the pump splines. Technician B says a pump support bracket should be added if over 40 lbs. Who is correct?
    A. A only
    B. B only
    C. Both A and B
    D. Neither A nor B   (B17)

79. The u-joint failed in the PTO (power take off) driveline. Technician A says the working angle was 3 degrees at the pump end. Technician B says the driveline could be too long. Who is correct?
   A. A only
   B. B only
   C. Both A and B
   D. Neither A nor B   (B17)

80. All of these are considered contaminants in a hydraulic system EXCEPT:
   A. Aeration
   B. Humidity
   C. Water
   D. Additives   (A1.1)

81. A hot shift PTO (power take off) lubrication line has been cut by road debris and is leaking heavily. Technician A says the PTO idler shaft could be damaged. Technician B says the input shaft could be damaged. Who is correct?
   A. A only
   B. B only
   C. Both A and B
   D. Neither A nor B   (B4)

82. The aluminum transmission housing has damage at the hot shift PTO (power take off) mounting surface. Technician A says the PTO could have experienced a hydraulic shock load. Technician B says the PTO could have been damaged by improper backlash. Who is correct?
   A. A only
   B. B only
   C. Both A and B
   D. Neither A nor B   (B17)

83. A loud clattering sound is heard in the transmission while operating a newly installed PTO (power take off). Technician A says a PTO gasket might need to be added. Technician B says a dial indicator will verify gear backlash. Who is correct?
   A. A only
   B. B only
   C. Both A and B
   D. Neither A nor B   (B14)

84. A driver operating a hydraulic jackhammer complains of very hot operation. Technician A says the oil cooler thermostat could be the cause. Technician B says the oil viscosity could be the cause. Who is correct?
   A. A only
   B. B only
   C. Both A and B
   D. Neither A nor B   (A6.3)

85. The operator complains of poor performance with a hot shift PTO (power take off). Technician A says the ATF (automatic transmission fluid) could be low. Technician B says the ATF might need to be serviced. Who is correct?
   A. A only
   B. B only
   C. Both A and B
   D. Neither A nor B   (B6)

86. The hydraulic system will not build pressure or flow with a hot shift PTO (power take off). Technician A says the PTO clutch plates could be worn. Technician B says the transmission shift pressure could be too high. Who is correct?
    A. A only
    B. B only
    C. Both A and B
    D. Neither A nor B   (B6)

87. Which of the connections is not considered a zero leak connection?
    A. JIC (Joint Industry Council)
    B. O-ring
    C. NPT (National Pipe Thread)
    D. Flat face   (A3.1)

88. When checking the hot shift PTO (power take off) system, the transmission oil is milky in color. Technician A says the transmission oil has been overheated and should be replaced. Technician B says the main shift pressure could be too high and will need to be adjusted. Who is correct?
    A. A only
    B. B only
    C. Both A and B
    D. Neither A nor B   (B6)

89. When performing a PM (Preventative Maintenance) check of a dump truck, a leak at the PTO (power take off) is noted at the mounting base. Technician A says to add fluid to top off the unit and service the unit at a later date. Technician B says the PTO gaskets might need to be replaced. Who is correct?
    A. A only
    B. B only
    C. Both A and B
    D. Neither A nor B   (B4)

90. A hot shift PTO (power take off) will not engage when activated. What is the most likely cause?
    A. Worn PTO fork
    B. Leaking shaft seal
    C. Kinked lube hose
    D. Worn clutch seal   (B7)

91. A hydraulic pump shows signs of cavitation damage. Technician A says a damaged tank diffuser could be the cause. Technician B says a damaged tank baffle could be the cause. Who is correct?
    A. A only
    B. B only
    C. Both A and B
    D. Neither A nor B   (A2.3)

92. A dump body will not fully lift when the shifter is held in the raise position. Technician A says the cable shifter might need adjustment. Technician B says the body pull-off cable might need adjustment. Who is correct?
    A. A only
    B. B only
    C. Both A and B
    D. Neither A nor B   (B5)

93. A wire shift PTO (power take off) fully engages, but pops out of gear under load. Technician A says the shifter cover could be loose. Technician B says the control cable end could be loose. Who is correct?
    A. A only
    B. B only
    C. Both A and B
    D. Neither A nor B   (B5)

94. A hydraulic pump was replaced but will not produce flow after being installed. Technician A says to crawl underneath the truck and verify that the driveline is rotating. Technician B says to increase the engine RPM (Revolutions per minute) to prime the pump. Who is correct?
    A. A only
    B. B only
    C. Both A and B
    D. Neither A nor B   (B8)

95. A PTO (power take off) is being installed. It will not fit on the RH (right-hand) side but it will fit on the LH (left-hand) side. Technician A says the output shaft rotation will be affected. Technician B says to check if the output shaft speed can be affected. Who is correct?
    A. A only
    B. B only
    C. Both A and B
    D. Neither A nor B   (B8)

96. While replacing a worn hot shift clutch pack seal, Technician A says to recover the ATF (automatic transmission fluid) and filter for reuse. Technician B says to reuse the lock tabs and transmission studs. Who is correct?
    A. A only
    B. B only
    C. Both A and B
    D. Neither A nor B   (B6)

97. A bi-rotational pump should be used with a:
    A. 4-way valve
    B. Winch motor
    C. Double-acting cylinder
    D. Hot shift PTO (power take off)   (A1.1)

98. A wire shift PTO (power take off) was converted to an air shift unit to allow the driver easier operation of the hydraulic circuit. Technician A says the operation decals should be changed in the cab. Technician B says the PTO warning labels should be updated. Who is correct?
    A. A only
    B. B only
    C. Both A and B
    D. Neither A nor B   (B9)

99. A hydraulic pump is believed to be starving during high load operation. Technician A says a flow meter should be used to diagnose the problem. Technician B says a collapsed inlet line could be the cause. Who is correct?
    A. A only
    B. B only
    C. Both A and B
    D. Neither A nor B   (A3.1)

100. When replacing a PTO (power take off) control cable due to corrosion, Technician A says to secure both ends of the cable to reduce free play. Technician B says not to exceed the minimum bend radius when installing the cable. Who is correct?
    A. A only
    B. B only
    C. Both A and B
    D. Neither A nor B   (B5)

101. A gear pump is being checked for loss of flow under load. Technician A says a pressure gauge should be used. Technician B says the ball valves should be closed to isolate the pump. Who is correct?
    A. A only
    B. B only
    C. Both A and B
    D. Neither A nor B   (A6.2)

102. All of these could cause driveline failure EXCEPT:
    A. Incorrect u-joint phasing
    B. Parallel U-joints
    C. 3-degree work angle
    D. Over greasing U-joints   (B10)

103. When reviewing the under clutch pump belt, the technician notices the V-Belt sunken into the pulley groove. The technician should:
    A. Reset the belt tension to specs
    B. Verify the belt deflection
    C. Replace the belt
    D. Apply belt dressing   (B13)

104. While testing a clutch pump hydraulic circuit, the system will not reach full system pressure. Technician A says a contaminated relief valve could be the cause. Technician B says a worn serpentine could be the cause. Who is correct?
    A. A only
    B. B only
    C. Both A and B
    D. Neither A nor B   (B13)

105. A PTO (power take off) driven rotary screw air compressor has been installed on a used utility truck. Technician A says a Warning Label should be installed to alert the driver of the engine RPM (revolutions per minute) Preset. Technician B says the operator of the unit should review the operation of the compressor with the mechanic. Who is correct?
    A. A only
    B. B only
    C. Both A and B
    D. Neither A nor B   (B9)

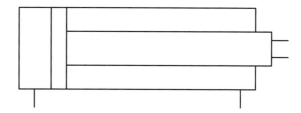

106. The symbol above represents a:
    A. Single-acting cylinder
    B. Double-acting cylinder
    C. 2-way manual selector valve
    D. 2-way flow control valve   (A5.6)

107. A crankshaft driven load sense pump generates a vibration during operation. Technician A says a worn pump shaft bushing could be the cause. Technician B says the PTO (power take off) lock tabs could be the cause. Who is correct?
    A. A only
    B. B only
    C. Both A and B
    D. Neither A nor B   (B14)

108. During the initial PTO (power take off) operation, a shudder is felt in the driveline. Technician A says the U-joints could be worn. Technician B says the PTO output shaft keyway could be worn. Who is correct?
    A. A only
    B. B only
    C. Both A and B
    D. Neither A nor B   (B14)

109. A hydraulically driven winch wire rope operates slowly at PTO (power take off) set speed but operates to specification once the throttle is depressed. Technician A says the parking brake interlock could be the cause. Technician B says the engine programming may need to be reset. Who is correct?
    A. A only
    B. B only
    C. Both A and B
    D. Neither A nor B   (B15)

110. The driveshaft for a crankshaft pump was removed for u-joint replacement. Technician A says the lock tight should not be applied to set screws to facilitate removal. Technician B says the driveline guards should be reinstalled to prevent any harm. Who is correct?
    A. A only
    B. B only
    C. Both A and B
    D. Neither A nor B   (B16)

111. The PTO (power take off) was mounted by the diesel particulate filter, due to obstruction on the driver side of the transmission. Technician A says the exhaust pipe might need to be relocated to prevent heat buildup. Technician B says a heat shield might need to be added to protect the PTO from exhaust pipes. Who is correct?
    A. A only
    B. B only
    C. Both A and B
    D. Neither A nor B   (B16)

112. During disassembly of a gear pump on a loader system, signs of scored wear plates were noticed. Technician A says excessive shaft RPM (revolutions per minute) could be the cause. Technician B says excessive oil temperatures could be the cause. Who is correct?
    A. A only
    B. B only
    C. Both A and B
    D. Neither A nor B   (A1.3)

113. The PTO (power take off) gears grind when engaging the wire control shifter. Technician A says the PTO could have a chipped gear. Technician B says the wire cable connector clamp could be loose. Who is correct?
    A. A only
    B. B only
    C. Both A and B
    D. Neither A nor B   (B14)

114. During inspection of the PTO (power take off), damage was noted to the driveline since the customer began using the vehicle off-road. Technician A says a skid plate might need to be added to protect the PTO. Technician B says the PTO might need to be relocated to another opening. Who is correct?
    A. A only
    B. B only
    C. Both A and B
    D. Neither A nor B   (B16)

115. The PTO (power take off) driveline failed and damaged the underneath of the chassis. Technician A says the slip joint should have been located on the pump side of the driveline. Technician B says a driveline hoop guard could have been installed to prevent damage. Who is correct?
    A. A only
    B. B only
    C. Both A and B
    D. Neither A nor B   (B16)

116. Technician A says the proper engine set speed can keep the hydraulic system from overheating. Technician B says the proper engine set speed can reduce the chances of pump cavitation. Who is correct?
    A. A only
    B. B only
    C. Both A and B
    D. Neither A nor B   (B15)

117. The pressure protection valve is used to:
    A. Operate the PTO (power take off) with the parking brakes on
    B. Control the air to the wet tank
    C. Limit the air pressure to the PTO
    D. Stop air leaks from affecting brakes   (C1)

118. Technician A says the pressure protection valve should be installed at the tank. Technician B says the pressure protection valve should be installed with the arrow point toward the PTO (power take off). Who is correct?
    A. A only
    B. B only
    C. Both A and B
    D. Neither A nor B   (C2)

119. The hydraulic system has been contaminated by water; all the filters have been changed and flushed out. Technician A says diesel fluid or a similar fluid could be used to wipe the inside of the oil tank. Technician B says depending on the level of contamination, the strainers in the tank should be replaced instead of cleaning. Who is correct?
   A. A only
   B. B only
   C. Both A and B
   D. Neither A nor B   (A2.2)

120. Technician A says mounting the PTO (power take off) hump down can change the PTO output rotation. Technician B says mounting the PTO hump up can reduce the PTO output speed. Who is correct?
   A. A only
   B. B only
   C. Both A and B
   D. Neither A nor B   (B8)

121. The rod end of the cylinder is leaking heavily. Technician A says it could be the rod wiper seal. Technician B says it could be the piston gland seal. Who is correct?
   A. A only
   B. B only
   C. Both A and B
   D. Neither A or B   (A5.3)

122. The hydraulic system has been contaminated with small pieces of rubber. Technician A says the hoses might not have been properly cleaned before installation. Technician B says the manufacturer might not have properly installed the pump shaft seal. Who is correct?
   A. A only
   B. B only
   C. Both A and B
   D. Neither A nor B   (A3.4)

123. The driver complains of sluggish hydraulic operation during the extreme cold weather. Technician A says the hydraulic system could be flushed and replaced with lower viscosity oil. Technician B says adding a heater to the oil tank could warm the oil prior to system operation. Who is correct?
   A. A only
   B. B only
   C. Both A and B
   D. Neither A nor B   (A6.4)

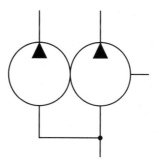

124. The symbol above represents a _____ pump.
   A. Dual flow
   B. 3- Line Dump
   C. Bi-rotational
   D. Tandem   (A1.1)

125. The fluid in a hydraulic circuit has been overheated and has a burnt odor. Technician A says the hydraulic oil should be replaced with a lower viscosity fluid. Technician B says the system duty cycle should be reviewed to ensure the operator did not exceed system design. Who is correct?
    A. A only
    B. B only
    C. Both A and B
    D. Neither A nor B   (A6.3)

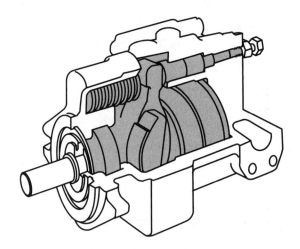

126. The performance of the component above is affected by all of these EXCEPT:
    A. Engine speed
    B. PTO (power take off) ratio
    C. Actuator speed
    D. Engine torque   (A1.1)

127. The driver complains that the PTO (power take off) will not engage into gear. Technician A says the PTO air line could be kinked. Technician B says the PTO shift cylinder could be damaged. Who is correct?
    A. A only
    B. B only
    C. Both A and B
    D. Neither A nor B   (C3)

128. Before installing the hose assembly with the JIC (Joint Industry Council) fittings into the hydraulic system, the mechanic should first:
    A. Apply Teflon tape
    B. Clean out the hose
    C. Lubricate the threads
    D. Lubricate the seal   (A3.4)

129. A direct mount pump is showing signs of damage at the mounting surface. Technician A says this could be a sign of excessive shaft speed. Technician B says rear pump brackets should be installed. Who is correct?
    A. A only
    B. B only
    C. Both A and B
    D. Neither A nor B   (A1.2)

130. When checking the oil level, it was discovered to have an algae like substance with a strong unpleasant odor. Technician A says the tank should be disinfected with chlorine. Technician B says the hydraulic oil should be removed and flushed. Who is correct?
  A. A only
  B. B only
  C. Both A and B
  D. Neither A nor B   (A2.3)

131. When reviewing the pump angle on a remote mounted pump the angle should be:
  A. The same as the frame rail
  B. The same as the transmission
  C. No more than 4 degrees
  D. Should be 0 degrees   (A1.2)

132. A newly assembled, single-acting dump truck generates a lot of turbulent flow inside the oil tank. Technician A says a larger oil tank could reduce the turbulence. Technician B says a larger inlet strainer could reduce the turbulence. Who is correct?
  A. A only
  B. B only
  C. Both A and B
  D. Neither A nor B   (A2.3)

133. After a hydraulic system was cleaned out and flushed due to contamination, the pump is noisy. Technician A says the pump should have been primed properly before starting. Technician B says a shop towel might have been left in the oil tank. Who is correct?
  A. A only
  B. B only
  C. Both A and B
  D. Neither A nor B   (A1.3)

134. Technician A says the air tubing used on a PTO (power take off) circuit should be DOT rated. Technician B says the air tubing should be routed away from heat sources. Who is correct?
  A. A only
  B. B only
  C. Both A and B
  D. Neither A nor B   (C3)

135. A direct-mounted pump has worn splines on the input shaft. Technician A says the pump's splines might not have been lubricated. Technician B says the engine harmonics could be the cause. Who is correct?
  A. A only
  B. B only
  C. Both A and B
  D. Neither A nor B   (A1.2, A1.3)

136. The hydraulic pump loses flow during operation. Technician A says a worn input shaft seal could be the cause. Technician B says the input shaft splines could be worn. Who is correct?
  A. A only
  B. B only
  C. Both A and B
  D. Neither A nor B   (A1.3)

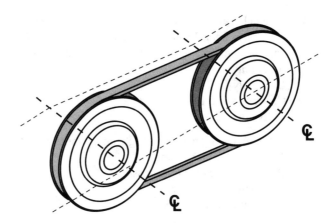

137. Technician A says the pulleys above can be checked with a straightedge bar. Technician B says the belt in the picture above can jump off pulley. Who is correct?
    A. A only
    B. B only
    C. Both A and B
    D. Neither A nor B   (B12)

138. During a systems check of a dump truck, the air operated hoist control does not feather the operation of the hoist. Technician A says a restricted air tube could be the cause. Technician B says the pressure regulator could be set too low. Who is correct?
    A. A only
    B. B only
    C. Both A and B
    D. Neither A nor B   (C3, C4)

139. The symbol shown above is a:
    A. Variable drive motor
    B. Air motor
    C. Directional check
    D. Check valve   (A6.1)

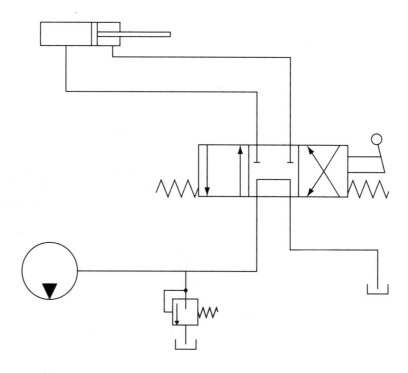

140. The circuit above operates slower and is running hotter than normal. What is the most likely cause?
    A. External pump leak
    B. Worn relief spring
    C. Pump over-speeding
    D. Restricted return filter   (A6.3)

141. While disassembling a gear pump, there is evidence of scoring type damage in the pump housing. What is the most likely cause?
    A. Torque converter slippage
    B. Constricted return filter
    C. Over pressurization
    D. Main relief stuck open   (A1.3)

142. An air leak is found at the pressure regulator for the PTO (power take off) single-acting shifter. Technician A says this could keep the PTO from shifting out of gear. Technician B says this could completely drain the truck's air tanks. Who is correct?
    A. A only
    B. B only
    C. Both A and B
    D. Neither A nor B   (C4)

143. Excessive pressure drop in a hydraulic system was found while diagnosing the hydraulic system. Technician A says the main relief valve could be opening too late. Technician B says this could raise the oil temperature. Who is correct?
    A. A only
    B. B only
    C. Both A and B
    D. Neither A nor B   (A6.3)

144. An underhood pump has been damaged due to cavitation. Before installing the new pump, Technician A says the mechanic needs to increase the suction hose diameter. Technician B says the mechanic needs to lower the oil tank mount. Who is correct?
    A. A only
    B. B only
    C. Both A and B
    D. Neither A nor B   (A1.3)

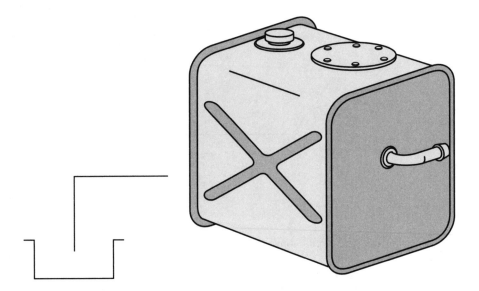

145. Technician A says the item displayed above should have a drain plug. Technician B says the item should be sized to fit the space available. Who is correct?
    A. A only
    B. B only
    C. Both A and B
    D. Neither A or B   (A1.2)

146. The hydraulic air compressor on a utility body is not operating smoothly. The compressor revs up and down constantly and will not produce rated air-flow. Technician A says the pump could have an internal leak. Technician B says the engine speed is not governing properly. Who is correct?
    A. A only
    B. B only
    C. Both A and B
    D. Neither A or B   (A1.3)

7 Appendices

Answers to the Test Questions for the Sample Test Section 5

1. C	16. B	31. A	46. B
2. A	17. C	32. A	47. A
3. C	18. A	33. B	48. C
4. C	19. B	34. C	49. B
5. C	20. B	35. C	50. C
6. C	21. C	36. A	
7. C	22. C	37. B	
8. A	23. C	38. D	
9. C	24. A	39. C	
10. B	25. B	40. C	
11. B	26. B	41. C	
12. D	27. C	42. C	
13. C	28. C	43. A	
14. B	29. B	44. C	
15. C	30. C	45. C	

Explanations to the Answers for the Sample Test Section 5

Question #1
Answer A is incorrect. Technician B is also correct.
Answer B is incorrect. Technician A is also correct.
Answer C is correct. Both Technicians are correct. A clogged strainer will starve the pump of oil supply and the relief valve could be set too low.
Answer D is incorrect. Both Technicians are correct.

Question #2
Answer A is correct. The noise is from the pump cavitation (aeration of oil).
Answer B is incorrect. The leak will not create a noise.
Answer C is incorrect. The driveline is correctly phased (an in-phase driveline will not cause noise).
Answer D is incorrect. A high relief setting will not create noise unless it's going over the threshold setting.

Question #3
 Answer A is incorrect. Technician B is also correct.
Answer B is incorrect. Technician A is also correct.
Answer C is correct. Both Technicians are correct. It is good practice to label the fill cap to prevent accidental contamination. Whenever changing a major component, it is also good practice to clean out the system.
Answer D is incorrect. Both Technicians are correct.

Question #4
Answer A is incorrect. Technician B is also correct.
Answer B is incorrect. Technician A is also correct.
Answer C is correct. Both Technicians are correct. The operating pressure and the direction of rotation are required to match the application.
Answer D is incorrect. Both Technicians are correct.

Question #5
Answer A is incorrect. Technician B is also correct.
Answer B is incorrect. Technician A is also correct.
Answer C is correct. Both Technicians are correct. The vacuum gauge will locate a suction leak that will cause pump fluid starvation and the strainer can be clogged, thus starving the pump of fluid.
Answer D is incorrect. Both Technicians are correct.

Question #6
Answer A is incorrect. Technician B is also correct.
Answer B is incorrect. Technician A is also correct.
Answer C is correct. Both Technicians are correct. The clutch pulleys should be checked with a straightedge to ensure these are in line. The burnishing of the clutch will allow the clutch to engage with more clamping force.
Answer D is incorrect. Both Technicians are correct.

Question #7
Answer A is incorrect. This is not a hydraulic pump symbol.
Answer B is incorrect. This is not a pneumatic pump symbol.
Answer C is correct. The symbol is of a hydraulic check valve.
Answer D is incorrect. This is not a relief valve symbol.

Question #8
Answer A is correct. A single-acting cylinder has only one working port, so air can be trapped in the piston side of the cylinder. A bleeder valve is usually ported into the cylinder to remove the air.
Answer B is incorrect. The rod seal has no function other than protect the cylinder rod.
Answer C is incorrect. Only Technician A is correct.
Answer D is incorrect. Only Technician A is correct.

Question #9
Answer A is incorrect. Technician B is also correct.
Answer B is incorrect. Technician A is also correct.
Answer C is correct. Both Technicians are correct. The leaking air tank could cause this issue, as well as the pressure protection valve.
Answer D is incorrect. Both Technicians are correct.

Question #10
Answer A is incorrect. The sticking valve explains the jerking movement but not the milky oil.
Answer B is correct. The milky oil appearance is an indication of aeration within the oil tank, which is the cause of the jerky operation (air is compressible).
Answer C is incorrect. On overfilled tank would not cause this failure.
Answer D is incorrect. A dirty return filter would not cause this failure.

Question #11
Answer A is incorrect. The oil filter should be changed.
Answer B is correct. The transmission oil does not need to be flushed.
Answer C is incorrect. The relief valve's operation should be checked; this may be the cause of the pump failure.
Answer D is incorrect. The hydraulic oil should be changed.

Question #12
Answer A is incorrect. Too many gaskets increase the backlash.
Answer B is incorrect. Excessive backlash would not cause this issue.
Answer C is incorrect. Neither Technician is correct.
Answer D is correct. Neither Technician is correct. This wear is typical of insufficient backlash.

Question #13
Answer A is incorrect. Pressure ratings do affect filter choice.
Answer B is incorrect. GPM ratings do affect filter choice.
Answer C is correct. This is a symbol for an oil filter. SSU is a measurement of oil, not a filter.
Answer D is incorrect. Micron rating does affect filter choice.

Question #14
Answer A is incorrect. A leaking filter will not affect operation.
Answer B is correct. A clogged breather will restrict the oil flow out of the tank (creating a vacuum) and cause a flow restriction to the pump.
Answer C is incorrect. An oversized inlet will not reduce flow to the pump.
Answer D is incorrect. Low oil viscosity will not affect pump flow.

Question #15
Answer A is incorrect. Technician B is also correct.
Answer B is incorrect. Technician A is also correct.
Answer C is correct. Both Technicians are correct. The low pressure or kinked hose can affect the PTO engagement.
Answer D is incorrect. Both Technicians are correct.

Question #16
Answer A is incorrect. JIC fittings, do not require sealant.
Answer B is correct. NPT fittings do require a thread sealant.
Answer C is incorrect. SAE fittings do not require a sealant.
Answer D is incorrect. UNF fittings do not require a sealant.

Question #17
Answer A is incorrect. Technician B is also correct.
Answer B is incorrect. Technician A is also correct.
Answer C is correct. Both Technicians are correct. The SAE rating should match the application and the diameter should match the application.
Answer D is incorrect. Both Technicians are correct.

Question #18
Answer A is correct. Only Technician A is correct. The driveline should be kept as short as possible.
Answer B is incorrect. The driveline should not be perpendicular to the frame, but parallel to the transmission.
Answer C is incorrect. Only Technician A is correct.
Answer D is incorrect. Technician A is correct.

Question #19
Answer A is incorrect. This rating is marked on the hose in PSI.
Answer B is correct. This refers to the inside diameter measurement of the hose.
Answer C is incorrect. The hose size is determined by the inside diameter of the hose.
Answer D is incorrect. This rating is marked on the hose in PSI.

Question #20
Answer A is incorrect. The valve is not a 3-way spool valve.
Answer B is correct. The valve is a 4-way, 3-section valve.
Answer C is incorrect. The valve shown is not a 2-position spool valve.
Answer D is incorrect. The valve shown is not a 2-position spool valve.

Question #21
Answer A is incorrect. Technician B is also correct.
Answer B is incorrect. Technician A is also correct.
Answer C is correct. Both Technicians are correct. The addition of the idler gear will alter the output rotation of the PTO and the pump rotation will need to be verified.
Answer D is incorrect. Both Technicians are correct.

Question #22
Answer A is incorrect. An oversized inlet line will not slow down the operation auger.
Answer B is incorrect. An excessive relief pressure setting will not affect the auger operation.
Answer C is correct. An excessive pressure drop will indicate an increase in the resistance within the circuit and the creation of excessive heat.
Answer D is incorrect. An oversized case drain will not affect the auger operational speed.

Question #23
Answer A is incorrect. Technician B is also correct.
Answer B is incorrect. Technician A is also correct.
Answer C is correct. Both Technicians are correct. A flow meter and a pressure gauge will help to diagnose a pump.
Answer D is incorrect. Both Technicians are correct.

Question #24
Answer A is correct. Only Technician A is correct. A bad crimp on the lube hose could restrict the oil to the PTO or cause a leak.
Answer B is incorrect. The restrictor or orifice would most likely cause a drop in lube line pressure in the transmission as well, but unlikely to cause this issue.
Answer C is incorrect. Only Technician A is correct.
Answer D is incorrect. Technician A is correct.

Question #25
Answer A is incorrect. Bleeding of the load sense line is performed as the system is operating.
Answer B is correct. Filling the housing with oil before starting the system will protect the pump from a dry starting condition (wear).
Answer C is incorrect. Setting the low pressure standby is performed on the system when it is in running mode.
Answer D is incorrect. The pump should not be run until the adjustments are performed.

Question #26
Answer A is incorrect. This configuration does not allow the body to rise faster.
Answer B is correct. Only Technician B is correct. This change alters the flow inside the pump to make it lower faster.
Answer C is incorrect. Only Technician B is correct.
Answer D is incorrect. Technician B is correct.

Question #27
Answer A is incorrect. Technician B is also correct.
Answer B is incorrect. Technician A is also correct.
Answer C is correct. Both Technicians are correct. A water separator can reduce moisture in the air lines; or the air line may be connected into the wet tank and should be relocated to the secondary tank.
Answer D is incorrect. Both Technicians are correct.

Question #28
Answer A is incorrect. The main relief valve is used to limit system operating pressure, not create pressure.
Answer B is incorrect. Pump displacement will not create load.
Answer C is correct. Any load on the system causes a restriction to flow. This creates pressure system pressure.
Answer D is incorrect. Cylinder displacement will not create load.

Question #29
Answer A is incorrect. This is not a 4-way section symbol.
Answer B is correct. This is a symbol for a relief valve.
Answer C is incorrect. This is not a check valve symbol.
Answer D is incorrect. This is not a flow divider symbol.

Question #30
Answer A is incorrect. The current backlash is correct.
Answer B is incorrect. The current backlash is correct.
Answer C is correct. The current backlash is correct and the lock tabs should be bent into place to retain the torque on the fasteners.
Answer D is incorrect. The transmission is a manual transmission and does not use ATF.

Question #31

Answer A is correct. Only Technician A is correct. Using the dye will help identify smaller, hard to find leaks.

Answer B is incorrect. The technicians should not use their (unprotected) hands, as fluid under high pressure can pierce the skin.

Answer C is incorrect. Only Technician A is correct.

Answer D is incorrect. Technician A is correct.

Question #32

Answer A is correct. A clogged case drain would cause this kind of leak by building up pressure within the pump housing.

Answer B is incorrect. A clogged filter would not create this leak.

Answer C is incorrect. A restricted inlet line would not create this leak.

Answer D is incorrect. A restricted inlet hose would create a cavitation effect.

Question #33

Answer A is incorrect. Feeler gauge is not the correct gauge to use.

Answer B is correct. Dial indicator is the proper instrument/gauge to use.

Answer C is incorrect. Slide calipers are not the correct instrument/gauge to use.

Answer D is incorrect. Depth gauge is not the correct gauge to use.

Question #34

Answer A is incorrect. It is not a measurement of flow capability.

Answer B is incorrect. It is not a beta measurement.

Answer C is correct. It is a measurement of filtering capability.

Answer D is incorrect. It is not a measurement of absorption.

Question #35

Answer A is incorrect. Technician A is also correct.

Answer B is incorrect. Technician B is also correct.

Answer C is correct. Both Technicians are correct. Contamination affects both filters and hydraulic oil and should be replaced.

Answer D is incorrect. Both Technicians are correct.

Question #36

Answer A is correct. Only Technician A is correct. The use of a solid shaft could cause this issue and perhaps should be changed to a standard (hollow) shaft.

Answer B is incorrect. The slip joint would not cause this issue (if properly lubricated).

Answer C is incorrect. Only Technician A is correct.

Answer D is incorrect. Technician A is correct.

Question #37

Answer A is incorrect. A hot shift PTO's backlash setting is determined by the gasket supplied and is not adjustable.

Answer B is correct. Only Technician B is correct. When restarting a system that had the pump replaced, it is good practice to back off the relief valves to prevent them from contamination or damage.

Answer C is incorrect. Only Technician B is correct.

Answer D is incorrect. Technician B is correct.

Question #38

Answer A is incorrect. The main relief protects the system and should not be adjusted without review of circuit.

Answer B is incorrect. The pump will not increase the load capacity.

Answer C is incorrect. Neither Technician is correct.

Answer D is correct. Neither Technician is correct.

Question #39
Answer A is incorrect. Technician B is also correct.
Answer B is incorrect. Technician A is also correct.
Answer C is correct. Both Technicians are correct. The PTO can be plumbed (located) into the wrong port, which would not allow the full pressure to engage the clutch pack. Also, if the shifter solenoid is wired/plumbed incorrectly, it also can cause this failure.
Answer D is incorrect. Both Technicians are correct.

Question #40
Answer A is incorrect. A leak on the rod end of the cylinder would not cause this drift condition.
Answer B is incorrect. The main relief setting would not cause this issue.
Answer C is correct. A sticking directional valve could cause this issue.
Answer D is incorrect. Worn pump gears would not cause this issue.

Question #41
Answer A is incorrect. The main relief valve would not cause this issue.
Answer B is incorrect. A clogged strainer would not cause this issue.
Answer C is correct. An overfilled reservoir is the cause of this issue.
Answer D is incorrect. The pump would not cause this issue.

Question #42
Answer A is incorrect. Technician B is also correct.
Answer B is incorrect. Technician A is also correct.
Answer C is correct. Both Technicians are correct. The driveline RPM and the working angles could be the cause.
Answer D is incorrect. Both Technicians are correct.

Question #43
Answer A is correct. The breather is not designed to prevent oil leaks.
Answer B is incorrect. The breather allows air pressure (atmospheric) to self-prime the pump.
Answer C is incorrect. The breather does not help keep moisture from the oil tank.
Answer D is incorrect. The breather does help exhaust the turbulent air/heat from the tank and allow the oil to release any air that might be trapped.

Question #44
Answer A is incorrect. An overfilled tank would not cause this issue
Answer B is incorrect. Clogged return filters would not cause this issue.
Answer C is correct. Incorrect hose fittings (angles) can create an excessive pressure drop, which may also create noise during high-flow operation.
Answer D is incorrect. A relief-valve set too high would not cause this issue.

Question #45
Answer A is incorrect. Technician B is also correct.
Answer B is incorrect. Technician A is also correct.
Answer C is correct. Both Technicians are correct. The set speed should be set to lock up RPM and the PTO ratio should be verified to ensure correct operation.
Answer D is incorrect. Both Technicians are correct.

Question #46
Answer A is incorrect. The RPM level will be set after performing the basic settings.
Answer B is correct. The relief should be backed off just in case the preset is too high.
Answer C is incorrect. The vent line would be a component of the manifold.
Answer D is incorrect. The air would be purged upon the system being cycled.

Question #47
Answer A is correct. The oil might have a slightly different viscosity so this could affect the pressure and flow.
Answer B is incorrect. While the oil is biodegradable, it still can be considered a safety hazard and needs to be cleaned up appropriately.
Answer C is incorrect. Only Technician A is correct.
Answer D is incorrect. Technician A is correct.

Question #48
Answer A is incorrect. Technician B is also correct.
Answer B is incorrect. Technician A is also correct.
Answer C is correct. Both Technicians are correct. A tandem pump will have two outlet pressure ports and can share a section inlet line if it's sized and ported correctly.
Answer D is incorrect. Both Technicians are correct.

Question #49
Answer A is incorrect. A single-acting cylinder will not have a port relief on the rod end.
Answer B is correct. The pump with excessive flow of oil under a load could stall the engine.
Answer C is incorrect. Only Technician B is correct.
Answer D is incorrect. Technician B is correct.

Question #50
Answer A is incorrect. Technician B is also correct.
Answer B is incorrect. Technician A is also correct.
Answer C is correct. The damaged line can cause the seal to fail and the failed seal or gaskets could create openings for outside contamination.
Answer D is incorrect. Both Technicians are correct.

Answers to the Test Questions for the Additional Test Questions Section 6

1. A	26. B	51. C	76. C
2. A	27. B	52. D	77. C
3. A	28. D	53. C	78. A
4. C	29. A	54. C	79. B
5. C	30. C	55. A	80. D
6. B	31. A	56. C	81. A
7. C	32. A	57. D	82. A
8. C	33. B	58. A	83. B
9. B	34. C	59. C	84. C
10. A	35. D	60. B	85. C
11. C	36. A	61. C	86. A
12. A	37. C	62. A	87. C
13. C	38. C	63. C	88. D
14. C	39. C	64. A	89. B
15. B	40. D	65. C	90. D
16. C	41. C	66. B	91. C
17. A	42. C	67. C	92. C
18. A	43. C	68. A	93. C
19. B	44. C	69. A	94. D
20. B	45. B	70. D	95. B
21. C	46. C	71. C	96. D
22. A	47. A	72. B	97. D
23. C	48. B	73. B	98. C
24. B	49. B	74. C	99. B
25. C	50. B	75. C	100. C

101. A	**113.** C	**125.** B	**137.** C
102. C	**114.** C	**126.** C	**138.** C
103. C	**115.** B	**127.** C	**139.** B
104. C	**116.** C	**128.** B	**140.** B
105. C	**117.** D	**129.** B	**141.** C
106. B	**118.** C	**130.** B	**142.** D
107. A	**119.** B	**131.** B	**143.** B
108. C	**120.** D	**132.** A	**144.** A
109. C	**121.** B	**133.** C	**145.** C
110. B	**122.** A	**134.** C	**146.** C
111. B	**123.** C	**135.** C	
112. C	**124.** D	**136.** B	

Explanations to the Answers for the Additional Test Questions Section 6

Question #1
Answer A is correct. Only Technician A is correct. A kinked hose to the rotation motor (restriction) could be the cause.
Answer B is incorrect. The main relief is not the cause because the issue is localized within the rotation.
Answer C is incorrect. Only Technician A is correct.
Answer D is incorrect. Technician A is correct.

Question #2
Answer A is correct. Load is determined by the system capacity.
Answer B is incorrect. The pump size will affect the system operating speed.
Answer C is incorrect. The pump size will affect the hose size.
Answer D is incorrect. The pump size will affect the valve size.

Question #3
Answer A is correct. Only Technician A is correct. A damaged seal could cause this issue.
Answer B is incorrect. This cylinder would not have a bleeder screw.
Answer C is incorrect. Only Technician A is correct.
Answer D is incorrect. Technician A is correct.

Question #4
Answer A is incorrect. Technician B is also correct.
Answer B is incorrect. Technician A is also correct.
Answer C is correct. Both Technicians are correct. Both a kink and incorrect fittings (size/angles) could cause excessive restriction within the circuit.
Answer D is incorrect. Both Technicians are correct.

Question #5
Answer A is incorrect. Technician B is also correct.
Answer B is incorrect. Technician A is also correct.
Answer C is correct. Both Technicians are correct. An improper relief valve setting or a worn pump seal could cause a drop in pressure and flow.
Answer D is incorrect. Both Technicians are correct.

Question #6
Answer A is incorrect. The main relief would not affect this issue.
Answer B is correct. Only Technician B is correct. The valve spool could cause this issue by not moving fully (sticking).
Answer C is incorrect. Only Technician B is correct.
Answer D is incorrect. Technician B is correct.

Question #7
Answer A is incorrect. Technician B is also correct.
Answer B is incorrect. Technician A is also correct.
Answer C is correct. Both Technicians are correct. Preventing a hose twist and the correct pressure rating will help reduce future hose failure.
Answer D is incorrect. Both Technicians are correct.

Question #8
Answer A is incorrect. Technician B is also correct.
Answer B is incorrect. Technician A is also correct.
Answer C is correct. Heat shields and an incorrect hose radius (kink/location) can be the cause of this failure.
Answer D is incorrect. Both Technicians are correct.

Question #9
Answer A is incorrect. The lubricant would not cause this issue.
Answer B is correct. This wear is an example of incorrect backlash (not set deep enough).
Answer C is incorrect. Insufficient backlash would not cause this issue.
Answer D is incorrect. An unbalanced driveline would not cause this issue.

Question #10
Answer A is correct. This is a RH helical cut gear.
Answer B is incorrect. This is not a LH helical cut gear.
Answer C is incorrect. This is not a spur gear.
Answer D is incorrect. This is not a sprocket.

Question #11
Answer A is incorrect. A broken gear would not cause this issue.
Answer B is incorrect. Excessive backlash would cause a clattering noise.
Answer C is correct. Insufficient backlash will cause this issue.
Answer D is incorrect. A broken shift fork would not cause this issue.

Question #12
Answer A is correct. The arrow symbol indicates a load sense compensating pump. The flow varies under system demand.
Answer B is incorrect. The arrow does not indicate adjustable pressure setting.
Answer C is incorrect. Only Technician A is correct.
Answer D is incorrect. Technician A is correct.

Question #13
Answer A is incorrect. Backlash would not cause this issue.
Answer B is incorrect. A broken gear would not cause this issue.
Answer C is correct. Unbalanced drive shaft could cause the seal to fail.
Answer D is incorrect. A worn PTO shifter would not cause this issue.

Question #14
Answer A is incorrect. Technician B is also correct.
Answer B is incorrect. Technician A is also correct.
Answer C is correct. Both Technicians are correct. Excessive shaft speed and out-of-balance condition could cause the wear.
Answer D is incorrect. Both Technicians are correct.

Question #15
Answer A is incorrect. This is not a gear pump.
Answer B is correct. This is a vane pump.
Answer C is incorrect. This is not a lobe pump.
Answer D is incorrect. This is not a clutch pump.

Question #16
Answer A is incorrect. Technician B is also correct.
Answer B is incorrect. Technician A is also correct.
Answer C is correct. Both Technicians are correct. Both the relief setting and the viscosity could lead to heat-related issues.
Answer D is incorrect. Both Technicians are correct.

Question #17
Answer A is correct. The pump might require a support bracket to keep it from torque-related movement on the mounting surface.
Answer B is incorrect. Gasket sealant should never be used on an automatic transmission.
Answer C is incorrect. Only Technician A is correct.
Answer D is incorrect. Technician A is correct.

Question #18
Answer A is correct. Only Technician A is correct. Liquid Teflon will properly seal the fitting.
Answer B is incorrect. Teflon tape should be avoided, if incorrectly installed it can create many contamination issues.
Answer C is incorrect. Only Technician A is correct.
Answer D is incorrect. Technician A is correct.

Question #19
Answer A is incorrect. The symbol is not a flow divider.
Answer B is correct. The symbol is an oil cooler.
Answer C is incorrect. The symbol is not a return filter.
Answer D is incorrect. The symbol is not a motor.

Question #20
Answer A is incorrect. The case drain would not cause this issue.
Answer B is correct. Only Technician B is correct. The hose for the load sense will delay the response time for the pump to react to the system demands.
Answer C is incorrect. Only Technician B is correct.
Answer D is incorrect. Technician B is correct.

Question #21
Answer A is incorrect. A bent rod would not be the cause.
Answer B is incorrect. The circuit does not have a port relief valve.
Answer C is correct. The valve could be sticking and the pump may not be able to exhaust sufficient flow through the relief valve.
Answer D is incorrect. A clogged return filter would not be the cause.

Question #22
Answer A is correct. Only Technician A is correct. The valve will control oil flow upstream of this item.
Answer B is incorrect. This valve will not set pressure.
Answer C is incorrect. Only Technician A is correct.
Answer D is incorrect. Technician A is correct.

Question #23
Answer A is incorrect. The symbol is not a counterbalance valve.
Answer B is incorrect. The symbol is not a 3-way port relief valve.
Answer C is correct. The symbol is a flow divider.
Answer D is incorrect. The symbol is not a shuttle valve.

Question #24
Answer A is incorrect. The replacement pump is probably not the issue.
Answer B is correct. Only Technician B is correct. The relief valve could be contaminated and will not fully seal.
Answer C is incorrect. Only Technician B is correct.
Answer D is incorrect. Technician B is correct.

Question #25
Answer A is incorrect. The symbol is not a dual check valve.
Answer B is incorrect. The symbol is not a flow divider.
Answer C is correct. The symbol is a shuttle valve.
Answer D is incorrect. The symbol is not a counterbalance valve.

Question #26
Answer A is incorrect. The main relief valve is not the issue.
Answer B is correct. Only Technician B is correct. An internal leak at the cylinder spool valve can cause the actuator to move.
Answer C is incorrect. Only Technician B is correct.
Answer D is incorrect. Technician B is correct.

Question #27
Answer A is incorrect. If the rod moves it will extend rather than retract.
Answer B is correct. Only Technician B is correct. The leaking seals could be causing this issue.
Answer C is incorrect. Only Technician B is correct.
Answer D is incorrect. Technician B is correct.

Question #28
Answer A is incorrect. Technician B is also correct.
Answer B is incorrect. Technician A is also correct.
Answer C is correct. Both Technicians are correct. This is a symbol of a hydraulic motor with a variable control setting.
Answer D is incorrect. Both Technicians are correct.

Question #29
Answer A is correct. Only Technician A is correct. The lubricant is placed on the spline area in order to reduce wear.
Answer B is incorrect. The pump should be less than 40 lbs.
Answer C is incorrect. Only Technician A is correct.
Answer D is incorrect. Technician A is correct.

Question #30
Answer A is incorrect. Technician B is also correct.
Answer B is incorrect. Technician A is also correct.
Answer C is correct. Both Technicians are correct. The pump rotation and the hoses installed in a reverse position (backwards) could cause this problem.
Answer D is incorrect. Both Technicians are correct.

Question #31
Answer A is correct. Only Technician A is correct. The valve is a 4-way, 3-section valve.
Answer B is incorrect. The valve is a motor spool.
Answer C is incorrect. Only Technician A is correct.
Answer D is incorrect. Technician A is correct.

Question #32
Answer A is correct. The valve is electric shift.
Answer B is incorrect. The relief valve is contained inside of the valve diagram.
Answer C is incorrect. The valve does have a power beyond port.
Answer D is incorrect. The valve does have a tank return port.

Question #33
Answer A is incorrect. The air will need to be removed by bleeding the cylinder.
Answer B is correct. The bleeder screw will need to be opened in order to remove the trapped air.
Answer C is incorrect. Only Technician B is correct.
Answer D is incorrect. Technician B is correct.

Question #34
Answer A is incorrect. Technician B is also correct.
Answer B is incorrect. Technician A is also correct.
Answer C is correct. Both Technicians are correct. The main internal relief will protect the valve and will operate as double-acting cylinders.
Answer D is incorrect. Both Technicians are correct.

Question #35
Answer A is incorrect. 100R1 is a return hose.
Answer B is incorrect. 100R2 is a pressure hose.
Answer C is incorrect. 100R3 is not a hydraulic hose.
Answer D is correct. 100R4 is an inlet suction hose.

Question #36
Answer A is correct. Only Technician A is correct. The gasket is designed to provide for the correct backlash setting.
Answer B is incorrect. Additional sealant should not be used on the gaskets (may change the backlash setting).
Answer C is incorrect. Only Technician A is correct.
Answer D is incorrect. Technician A is correct.

Question #37
Answer A is incorrect. Technician B is also correct.
Answer B is incorrect. Technician A is also correct.
Answer C is correct. Both Technicians are correct. A defective main relief spring could cause overheating of the oil (resulting in premature lubricant failure) and may cause additional system damage.
Answer D is incorrect. Both Technicians are correct.

Question #38
Answer A is incorrect. Technician B is also correct.
Answer B is incorrect. Technician A is also correct.
Answer C is correct. Both Technicians are correct. The motor could be worn and an internal leak could occur. The damaged relief valve seat could also cause oil to relieve below the relief pressure-setting.
Answer D is incorrect. Both Technicians are correct.

Question #39
Answer A is incorrect. Technician B is also correct.
Answer B is incorrect. Technician A is also correct.
Answer C is correct. Both Technicians are correct. The oil should be replaced and if the circuit has exceeded recommended operating temperatures, it may have affected the springs in the relief valve.
Answer D is incorrect. Both Technicians are correct.

Question #40
Answer A is incorrect. A leaking return filter is not the issue.
Answer B is incorrect. A clogged tank breather is not the issue.
Answer C is incorrect. A worn filter bypass spring is not the issue.
Answer D is correct. Oil viscosity (too thick) could create an excessive pressure drop.

Question #41
Answer A is incorrect. Technician B is also correct.
Answer B is incorrect. Technician A is also correct.
Answer C is correct. Both Technicians are correct. Keeping the hoses as clean as possible and correct routing (away from sharp edges) will help maintain correct system operation.
Answer D is incorrect. Both Technicians are correct.

Question #42
Answer A is incorrect. Technician B is also correct.
Answer B is incorrect. Technician A is also correct.
Answer C is correct. Both Technicians are correct. Shock loads could cause this type of damage and the filters in bypass mode can indicate that contaminants are present within the system.
Answer D is incorrect. Both Technicians are correct.

Question #43
Answer A is incorrect. Technician B is also correct.
Answer B is incorrect. Technician A is also correct.
Answer C is correct. Both Technicians are correct. The driveline speed and balance can cause both PTO and input shaft leak.
Answer D is incorrect. Both Technicians are correct.

Question #44
Answer A is incorrect. Technician B is also correct.
Answer B is incorrect. Technician A is also correct.
Answer C is correct. Both Technicians are correct. The hydraulic system should begin initial operations, without a load, and all the actuators should be operated.
Answer D is incorrect. Both Technicians are correct.

Question #45
Answer A is incorrect. The oil condition will not affect the shaft spline area.
Answer B is correct. Only Technician B is correct. Excessive or shock loading can cause this issue.
Answer C is incorrect. Only Technician B is correct.
Answer D is incorrect. Technician B is correct.

Question #46
Answer A is incorrect. Technician B is also correct.
Answer B is incorrect. Technician A is also correct.
Answer C is correct. Both Technicians are correct. Worn relief spring tension or a bent rod can cause this issue.
Answer D is incorrect. Both Technicians are correct.

Question #47
Answer A is correct. Only Technician A is correct. A blocked section beyond the valve can cause this failure as this oil is directly from the pump and typically is not protected by the main relief.
Answer B is incorrect. A stuck-open relief valve would not cause this issue.
Answer C is incorrect. Only Technician A is correct.
Answer D is incorrect. Technician A is correct.

Question #48
Answer A is incorrect. The main relief setting should be above (higher than) the port relief setting; otherwise the port relief will not operate properly.
Answer B is correct. Only Technician B is correct. The relief valves should be set at operating temperature in order to achieve correct valve setting.
Answer C is incorrect. Only Technician B is correct.
Answer D is incorrect. Technician B is correct

Question #49
Answer A is incorrect. Length of hose does not affect or keep the pump primed.
Answer B is correct. Only Technician B is correct. The pump could be losing its prime overnight due to the tank being mounted lower than the pump (drain-back). By raising the oil tank, it will help keep the pump primed with oil so it won't cavitate.
Answer C is incorrect. Only Technician B is correct.
Answer D is incorrect. Technician B is correct.

Question #50
Answer A is incorrect. An oversized tank will help reduce turbulence.
Answer B is correct. Low oil level can cause aeration, as there may be insufficient oil volume to operate at the desired temperature, or the return flow point may be above the oil level and allow air to enter the tank above the fluid line and cause turbulence (aeration).
Answer C is incorrect. A restricted inlet screen would not be the cause.
Answer D is incorrect. An excessive pressure setting would not be the cause.

Question #51
Answer A is incorrect. Technician B is also correct.
Answer B is incorrect. Technician A is also correct.
Answer C is correct. Both Technicians are correct. An incorrect fitting or hose size can create excessive pressure drop, which translates into heat and poor performance.
Answer D is worn. Both Technicians are correct.

Question #52
Answer A is incorrect. The tank should have a fill screen.
Answer B is incorrect. The tank should have a magnet plug.
Answer C is incorrect. The tank should have a clean-out cover.
Answer D is correct. The tank should not have a desiccant filter.

Question #53
Answer A is incorrect. Technician B is also correct.
Answer B is incorrect. Technician A is also correct.
Answer C is correct. Both Technicians are correct. The pressure and return hoses should be separated from each other and away from sharp edges. Pressure and return hoses operate at different rates of pressure and flow; and if bundled together, they can create abrasion wear upon each other.
Answer D is incorrect. Both Technicians are correct.

Question #54
Answer A is incorrect. Technician B is also correct.
Answer B is incorrect. Technician A is also correct.
Answer C is correct. Both Technicians are correct. The main relief and the low oil level can affect the movement of the actuator.
Answer D is incorrect. Both Technicians are correct.

Question #55
Answer A is correct. Only Technician A is correct. The oil viscosity (resistance-to-flow) could be incorrect, which could cause oil breakdown under extended use operation (check duty cycle operation).
Answer B is incorrect. There are other methods to control this heat buildup than by telling the operator to not use his truck. Also, the duty cycle of the unit should be reviewed.
Answer C is incorrect. Only Technician A is correct.
Answer D is incorrect. Technician A is correct.

Question #56
Answer A is incorrect. Technician B is also correct.
Answer B is incorrect. Technician A is also correct.
Answer C is correct. Both Technicians are correct. Warning labels need to be located in the appropriate locations. The technician should not go under the truck with the driveline operational (not turning during servicing), in order to observe driveline integrity.
Answer D is incorrect. Both Technicians are correct.

Question #57
Answer A is incorrect. Its main purpose is not to reduce vibration to transmission.
Answer B is incorrect. It should be mounted on the PTO end in case it breaks.
Answer C is incorrect. It is not intended to support the center bearing.
Answer D is correct. Its main purpose is to allow the length of the shaft to change.

Question #58
Answer A is correct. Only Technician A is correct. Using more than three gaskets can loosen up over time and create leaks and damage to the mounting of the PTO.
Answer B is incorrect. The backlash should be between 0.008" to 0.012".
Answer C is incorrect. Only Technician A is correct.
Answer D is incorrect. Technician A is correct.

Question #59
Answer A is incorrect. Improper PTO backlash is not the issue.
Answer B is incorrect. A too low PTO ratio would not be the heat issue.
Answer C is correct. The PTO ratio can be too high and result in the output shaft to operate at a higher speed. The technician needs to review the replacement PTO application.
Answer D is incorrect. Improper PTO oil viscosity is not the issue.

Question #60
Answer A is incorrect. The item is not measured in BTUs.
Answer B is correct. Only Technician B is correct. The item is an oil filter that uses microns as one of the factors in determining its ability to remove particles, and is measured in microns.
Answer C is incorrect. Only Technician B is correct.
Answer D is incorrect. Technician B is correct

Question #61
Answer A is incorrect. Technician B is also correct.
Answer B is incorrect. Technician A is also correct.
Answer C is correct. Both Technicians are correct. The crack could be caused by excessive backlash or the lock tabs being loose, allowing the movement of the PTO.
Answer D is incorrect. Both Technicians are correct.

Question #62
Answer A is correct. Only Technician A is correct. The low torque converter pressure could cause the plates to slip and wear prematurely.
Answer B is incorrect. The lube line is not the issue.
Answer C is incorrect. Only Technician A is correct.
Answer D is incorrect. Technician A is correct.

Question #63
Answer A is incorrect. Technician B is also correct.
Answer B is incorrect. Technician A is also correct.
Answer C is correct. Both Technicians are correct. The shift hose could reduce the pressure to lock in the PTO clutch plates and the incorrect set speed can be too low to place the torque converter into the lock-up mode.
Answer D is incorrect. Both Technicians are correct.

Question #64
Answer A is correct. Only Technician A is correct. A dirty (plugged) or damaged oil cooler can allow the transmission to overheat and create foaming (aeration over-heat) within the transmission.
Answer B is incorrect. The gear ratio is not the issue.
Answer C is incorrect. Only Technician A is correct.
Answer D is incorrect. Technician A is correct.

Question #65
Answer A is incorrect. Technician B is also correct.
Answer B is incorrect. Technician A is also correct.
Answer C is correct. Both Technicians are correct. A loose shifter cable or the clutch pedal adjustment can affect the PTO engagement.
Answer D is incorrect. Both Technicians are correct.

Question #66
Answer A is incorrect. The exhaust system will affect the PTO mounting location.
Answer B is correct. The pump rotation will not be affected by the mounting location.
Answer C is incorrect. The frame clearance will affect the PTO mounting location
Answer D is incorrect. The gear ratio can vary depending on the size of the transmission chosen.

Question #67
Answer A is incorrect. Technician B is also correct.
Answer B is incorrect. Technician A is also correct.
Answer C is correct. Both Technicians are correct. The loose set screws and the slip joint can cause the driveline to fall off the PTO shaft depending on the amount of vibration, speed, and movement of the driveline.
Answer D is incorrect. Both Technicians are correct.

Question #68
Answer A is correct. The driveline is properly assembled.
Answer B is incorrect. The driveline is not phased correctly.
Answer C is incorrect. The slip joint and the phasing are not correct.
Answer D is incorrect. The slip joint is not on the PTO side.

Question #69
Answer A is correct. The gear is a LH helical cut.
Answer B is incorrect. The gear is not a hypoid.
Answer C is incorrect. The gear is not a RH helical cut.
Answer D is incorrect. The gear is not a spur cut.

Question #70
Answer A is incorrect. The pump resistance is not the issue.
Answer B is incorrect. The backlash is correct and is not the issue.
Answer C is incorrect. Neither Technician is correct.
Answer D is correct. Neither Technician is correct.

Question #71
Answer A is incorrect. A pressure gauge will not measure pump flow.
Answer B is incorrect. The pressure gauges are not measuring the main relief setting in this scenario.
Answer C is correct. By comparing the two gauges the technician can verify if there is a restriction between the two points and pressure drop.
Answer D is incorrect. The gauges are not plumbed into the load sense line.

Question #72
Answer A is incorrect. Belt misalignment is not the issue.
Answer B is correct. The belt is too tight and can cause bearing failure.
Answer C is incorrect. A worn belt tension unit is not the issue.
Answer D is incorrect. Worn pulley grooves are not the issue.

Question #73
Answer A is incorrect. The manufacturer will instruct when the lube line is needed.
Answer B is correct. Only Technician B is correct. The line should be rated to the pressure rating of the circuit.
Answer C is incorrect. Only Technician B is correct.
Answer D is incorrect. Technician B is correct.

Question #74
Answer A is incorrect. Technician B is also correct.
Answer B is incorrect. Technician A is also correct.
Answer C is correct. Both Technicians are correct. Improper u-joint angles can create vibrations or wobble that damage the seals and incorrect grease seals can also lead to damage of the u-joints.
Answer D is incorrect. Both Technicians are correct.

Question #75
Answer A is incorrect. Technician B is also correct.
Answer B is incorrect. Technician A is also correct.
Answer C is correct. Both Technicians are correct. A loose belt can slip under load or jump off the pulley groove(s).
Answer D is incorrect. Both Technicians are correct.

Question #76
Answer A is incorrect. Technician B is also correct.
Answer B is incorrect. Technician A is also correct.
Answer C is correct. Both Technicians are correct. The vibration can be due to driveline phasing or out-of-balance issues.
Answer D is incorrect. Both Technicians are correct.

Question #77
Answer A is incorrect. Technician B is also correct.
Answer B is incorrect. Technician A is also correct.
Answer C is correct. Both Technicians are correct. The correct PTO set speeds ensure that the torque converter and the PTO output shaft speed will perform at an optimum setting.
Answer D is incorrect. Both Technicians are correct.

Question #78
Answer A is correct. Only Technician A is correct. The splines should be lubricated on a direct mount PTO.
Answer B is incorrect. The pump should be remote mounted when over 50 lbs.
Answer C is incorrect. Only Technician A is correct.
Answer D is incorrect. Technician A is correct.

Question #79
Answer A is incorrect. The driveline u-joint at a 3-degree working angle would cause the failure.
Answer B is correct. Only Technician B is correct. The driveline length could be too long and create this issue.
Answer C is incorrect. Only Technician B is correct.
Answer D is incorrect. Technician B is correct.

Question #80
Answer A is incorrect. Aeration is considered harmful to the circuit components.
Answer B is incorrect. Humidity can create water within the circuit, which is harmful to components.
Answer C is incorrect. Water is considered harmful to the circuit components.
Answer D is correct. Additives are added to the hydraulic oil to help reduce foaming and moisture absorption.

Question #81
Answer A is correct. Only Technician A is correct. The loss of the lube line oil could create damage to the PTO shaft and should be checked.
Answer B is incorrect. The input shaft would not be affected.
Answer C is incorrect. Only Technician A is correct.
Answer D is incorrect. Technician A is correct.

Question #82
Answer A is correct. Only Technician A is correct. The transmission could have experienced a shock load and damaged the mounting.
Answer B is incorrect. The backlash should not be an issue as the hot shift PTO has a preset gasket for backlash.
Answer C is incorrect. Only Technician A is correct.
Answer D is incorrect. Technician A is correct.

Question #83
Answer A is incorrect. Adding a gasket might make this clattering worse if the backlash is the issue.
Answer B is correct. The backlash amount appears to be too great and should be checked with a dial indicator.
Answer C is incorrect. Technician B is correct.
Answer D is incorrect. Technician B is correct.

Question #84
Answer A is incorrect. Technician B is also correct.
Answer B is incorrect. Technician A is also correct.
Answer C is correct. Both Technicians are correct. A malfunctioning thermostat and the incorrect oil viscosity can affect oil and jackhammer operating temperature.
Answer D is incorrect. Both Technicians are correct.

Question #85
Answer A is incorrect. Technician B is also correct.
Answer B is incorrect. Technician A is also correct
Answer C is correct. Both Technicians are correct. The ATF (automatic transmission fluid) type and condition will affect PTO operation, as well as transmission operation, and needs to be checked.
Answer D is incorrect. Both Technicians are correct.

Question #86
Answer A is correct. Only Technician A is correct. The PTO or the transmission clutch plates can be worn due to overuse or other wear items.
Answer B is incorrect. It is unlikely that the shift pressure is too high. Typically it is more of an issue when it is too low.
Answer C is incorrect. Only Technician A is correct.
Answer D is incorrect. Technician A is correct.

Question #87
Answer A is incorrect. A JIC fitting is considered a high-pressure zero leak fitting.
Answer B is incorrect. An O-ring fitting is considered a high-pressure zero leak fitting.
Answer C is correct. A NPT fitting is older style fitting, which needs an added sealant to properly complete the connection.
Answer D is incorrect. A flat-face fitting is considered a high-pressure zero leak fitting.

Question #88
Answer A is incorrect. If the oil was overheated it probably would be dark in color and have a varnished look and odor.
Answer B is incorrect. The shift pressure is unlikely; more likely due to shock load or overuse.
Answer C is incorrect. Neither Technician is correct.
Answer D is correct. Neither Technician is correct. The oil appears to have been contaminated by water or another fluid. It could be due to turbulence but the shift pressure is probably not the issue.

Question #89
Answer A is incorrect. Ideally the unit should be repaired immediately. It is during a PM check that these items should be handled.
Answer B is correct. Only Technician B is correct. The leak is probably due to the gaskets, but should be verified.
Answer C is incorrect. Only Technician B is correct.
Answer D is incorrect. Technician B is correct.

Question #90
Answer A is incorrect. A hot shift PTO normally does not have a shift fork.
Answer B is incorrect. A leaking shaft seal is unlikely to cause this issue.
Answer C is incorrect. A kinked lube hose is unlikely to cause this issue.
Answer D is correct. A worn or damaged clutch pack seal could cause this issue.

Question #91
Answer A is incorrect. Technician B is also correct.
Answer B is incorrect. Technician A is also correct.
Answer C is correct. Both Technicians are correct. The damaged diffuser and baffle can result in reduced ability to control the turbulence within the oil tank.
Answer D is incorrect. Both Technicians are correct.

Question #92
Answer A is incorrect. Technician B is also correct
Answer B is incorrect. Technician A is also correct.
Answer C is correct. Both Technicians are correct. The cable shifter for the hoist and the pull off cable can affect the operation and might need to be adjusted.
Answer D is incorrect. Both Technicians are correct.

Question #93
Answer A is incorrect. Technician B is also correct
Answer B is incorrect. Technician A is also correct.
Answer C is correct. Both Technicians are correct. The shifter cover and cable could be loose.
Answer D is incorrect. Both Technicians are correct.

Question #94
Answer A is incorrect. The technician, for his safety, should not go under the truck while it is running.
Answer B is incorrect. The technician should not increase the RPM until the hydraulic system has been primed properly or it can damage the pump (cavitation).
Answer C is incorrect. Neither Technician is correct.
Answer D is correct. Neither Technician is correct.

Question #95
Answer A is incorrect. The output shaft rotation will not be affected unless an adapter gear is installed also.
Answer B is correct. Only Technician B is correct. The ratio could change depending on the side of the transmission the PTO is mounted. The technician should refer to the PTO catalog/transmission specifications for additional information.
Answer C is incorrect. Only Technician B is correct.
Answer D is incorrect. Technician B is correct.

Question #96
Answer A is incorrect. It is not a good practice to reuse ATF fluid.
Answer B is incorrect. It is a good practice not to reuse hardware that has been torqued or lock tabs that have been previously used.
Answer C is incorrect. Neither Technician is correct.
Answer D is correct. Neither Technician is correct.

Question #97
Answer A is incorrect. A 4-valve operation does not require a special pump.
Answer B is incorrect. A winch motor operation does not require a special pump.
Answer C is incorrect. The double-acting cylinder operation does not require a special pump.
Answer D is correct. The Hot Shift PTO runs CW and a bi-rotational or a W-pump should be used.

Question #98
Answer A is incorrect. Technician B is also correct.
Answer B is incorrect. Technician A is also correct.
Answer C is correct. Both Technicians are correct. All the operation decals and warning labels
should be updated.
Answer D is incorrect. Both Technicians are correct.

Question #99
Answer A is incorrect. Only Technician B is correct. The flow meter would not be used on the suction
side to measure flow.
Answer B is correct. A collapsed inlet hose would starve the pump of fluid flow.
Answer C is incorrect. Only Technician B is correct.
Answer D is incorrect. Technician B is correct.

Question #100
Answer A is incorrect. Technician B is also correct.
Answer B is incorrect. Technician A is also correct.
Answer C is correct. Both Technicians are correct. The PTO wire shift cable clamps could be loose
and the bend radius could also be the cause of this issue.
Answer D is incorrect. Both Technicians are correct.

Question #101
Answer A is correct. A pressure gauge will help diagnose a faulty pump.
Answer B is incorrect. A ball valve will allow isolation of the pump for removal.
Answer C is incorrect. Only Technician A is correct.
Answer D is incorrect. Technician A is correct.

Question #102
Answer A is incorrect. Incorrect u-joint phasing can cause driveline failure.
Answer B is incorrect. Parallel u-joints can cause driveline failure.
Answer C is correct. A 3-degree work angle will not cause this failure.
Answer D is incorrect. Over-greasing u-joints can cause driveline failure, as seals may become
damaged.

Question #103
Answer A is incorrect. Resetting the belt tension to specs will not correct this issue.
Answer B is incorrect. The belt is probably worn due to over-tightening.
Answer C is correct. Replacing the belt is the appropriate action.
Answer D is incorrect. Applying belt dressing will not resolve this issue.

Question #104
Answer A is incorrect. Technician B is also correct.
Answer B is incorrect. Technician A is also correct.
Answer C is correct. Both Technicians are correct. A contaminated relief valve will not fully close or
a slipping belt can create this issue.
Answer D is incorrect. Both Technicians are correct.

Question #105
Answer A is incorrect. Technician B is also correct.
Answer B is incorrect. Technician A is also correct.
Answer C is correct. Both Technicians are correct. Labels should be updated and the operation
procedures should be reviewed with the operator.
Answer D is incorrect. Both Technicians are correct.

Question #106
Answer A is incorrect. This is not a single-acting cylinder.
Answer B is correct. The diagram is a double-acting cylinder.
Answer C is incorrect. This is not a selector valve.
Answer D is incorrect. This is not a flow control valve.

Question #107
Answer A is correct. Only Technician A is correct. The pump bushing could be worn and cause the shaft to wobble and generate a vibration.
Answer B is incorrect. The PTO lock tabs are not the issue.
Answer C is incorrect. Only Technician A is correct.
Answer D is incorrect. Technician A is correct.

Question #108
Answer A is incorrect. Technician B is also correct.
Answer B is incorrect. Technician A is also correct.
Answer C is correct. Both Technicians are correct. The u-joints and the keyway within the input shaft could be worn and cause a shudder.
Answer D is incorrect. Both Technicians are correct.

Question #109
Answer A is incorrect. Technician B is also correct.
Answer B is incorrect. Technician A is also correct.
Answer C is correct. The PTO operational set speed is probably set too low. Most PTO set speeds require that there be an interlock to keep the truck from moving in this mode.
Answer D is incorrect. Both Technicians are correct.

Question #110
Answer A is incorrect. The set screws should be tightened down to ensure they do not vibrate loose. A compound such as "Loc-tite" is commonly used to ensure that fasteners will not loosen.
Answer B is correct. Only Technician B is correct. Any safety guards should be reinstalled.
Answer C is incorrect. Only Technician B is correct.
Answer D is incorrect. Technician B is correct.

Question #111
Answer A is incorrect. The DPF cannot be relocated to make room for the PTO.
Answer B is correct. Only Technician B is correct. Exhaust shields should be added whenever a PTO is in contact or close proximity with heat sources.
Answer C is incorrect. Only Technician B is correct.
Answer D is incorrect. Technician B is correct.

Question #112
Answer A is incorrect. Technician B is also correct.
Answer B is incorrect. Technician A is also correct.
Answer C is correct. An excessive output shaft RPM and oil temperature can contribute to wear in the plates.
Answer D is incorrect. Both Technicians are correct.

Question #113
Answer A is incorrect. Technician B is also correct.
Answer B is incorrect. Technician A is also correct.
Answer C is correct. Both Technicians are correct. A chipped gear or a loose cable connector could be the issue.
Answer D is incorrect. Both Technicians are correct.

Question #114
Answer A is incorrect. Technician B is also correct.
Answer B is incorrect. Technician A is also correct.
Answer C is correct. Both Technicians are correct. A skid plate might help protect the PTO and driveline or the technician could review and relocate to another PTO mounting location in order to add ground clearance.
Answer D is incorrect. Both Technicians are correct.

Question #115
Answer A is incorrect. The slip joint needs to be located at the PTO end.
Answer B is correct. Only Technician B is correct. A driveline hoop guard could help limit the damage caused by a broken driveline.
Answer C is incorrect. Only Technician B is correct.
Answer D is incorrect. Technician B is correct.

Question #116
Answer A is incorrect. Technician B is also correct.
Answer B is incorrect. Technician A is also correct.
Answer C is correct. Both Technicians are correct. The proper set speed can keep the pump from over-heating and reduce the chance of damaging the PTO due to incorrect engine RPM.
Answer D is incorrect. Both Technicians are correct.

Question #117
Answer A is incorrect. It is not intended to be a component of PTO operation with or without the parking brakes on.
Answer B is incorrect. Is not intended to control the air to the wet tank.
Answer C is incorrect. Is not intended to limit the air pressure to the PTO.
Answer D is correct. Pressure protection valves (single/double check) are designed to prevent air leaks from affecting brakes by limiting the air passing through the switch until a PSI threshold is reached.

Question #118
Answer A is incorrect. Technician B is also correct.
Answer B is incorrect. Technician A is also correct.
Answer C is correct. Both Technicians are correct. The valve should be mounted directly to the air tank and the flow through should flow in the direction of the PTO air circuit.
Answer D is incorrect. Both Technicians are correct.

Question #119
Answer A is incorrect. Only clean out the system with a non-corrosive fluid.
Answer B is correct. Only Technician B is correct. The strainers might need to be cleaned out or replaced depending on the level of contamination.
Answer C is incorrect. Only Technician B is correct.
Answer D is incorrect. Technician B is correct.

Question #120
Answer A is incorrect. The mounting of the PTO hump down will not alter rotation.
Answer B is incorrect. The mounting of the PTO hump up will not alter the PTO speed.
Answer C is incorrect. Neither Technician is correct.
Answer D is correct. Neither Technician is correct.

Question #121
Answer A is incorrect. The wiper seal keeps dirt/material off of the rod.
Answer B is correct. Only Technician B is correct. The piston gland could be the cause.
Answer C is incorrect. Only Technician B is correct.
Answer D is incorrect. Technician B is correct.

Question #122
Answer A is correct. Only Technician A is correct. All hoses should be cleaned before installation.
Answer B is incorrect. The pump seal would not cause this issue.
Answer C is incorrect. Only Technician A is correct.
Answer D is incorrect. Technician A is correct.

Question #123
Answer A is incorrect. Technician B is also correct.
Answer B is incorrect. Technician A is correct.
Answer C is correct. Both Technicians are correct. Changing the oil to lower viscosity can assist in system operation. Also, flushing the system can remove any moisture trapped in the oil. An oil tank warmer could also be a viable option.
Answer D is incorrect. Both Technicians are correct.

Question #124
Answer A is incorrect. It is not a symbol that indicates a dual flow circuit.
Answer B is incorrect. It is not indicating a 3-line pump configuration.
Answer C is incorrect. The pump symbols displayed are not bi-rotational.
Answer D is correct. The symbol represents a tandem pump.

Question #125
Answer A is incorrect. The oil viscosity rating may need to be changed due to overheating but the system design should be reviewed to see if something else could be the issue.
Answer B is correct. Only Technician B is correct. The duty cycle should be reviewed as part of the diagnostic process.
Answer C is incorrect. Only Technician B is correct.
Answer D is incorrect. Technician B is correct.

Question #126
Answer A is incorrect. The engine speed will affect the pump operation.
Answer B is incorrect. The PTO ratio will affect the pump operation.
Answer C is correct. The pump does not affect actuator speed.
Answer D is incorrect. Engine torque will affect the pump operation.

Question #127
Answer A is incorrect. Technician B is also correct.
Answer B is incorrect. Technician A is also correct
Answer C is correct. Both Technicians are correct. The PTO shift operation will be affected by a kink in the air line circuit. If the PTO cylinder is damaged it will also affect PTO operation.
Answer D is incorrect. Both Technicians both correct.

Question #128
Answer A is incorrect. Teflon tape is not used on the JIC fitting.
Answer B is correct. The hose should be cleaned out of any debris.
Answer C is incorrect. Lubrication is not needed on the JIC fitting.
Answer D is incorrect. There is no seal to lubricate.

Question #129
Answer A is incorrect. The shaft speed would not affect the mounting surface; typically torque on the pump would create this.
Answer B is correct. Only Technician B is correct. A rear pump bracket would help support the pump from torque-induced movement.
Answer C is incorrect. Only Technician B is correct.
Answer D is incorrect. Technician B is correct.

Question #130
Answer A is incorrect. The tank should not be cleaned with chlorine.
Answer B is correct. Only Technician B is correct. The tank should be flushed and refilled with oil.
Answer C is incorrect. Only Technician B is correct.
Answer D is incorrect. Technician B is correct.

Question #131
Answer A is incorrect. The pump should not be parallel with the frame rail.
Answer B is correct. The pump angle should be the same as the transmission.
Answer C is incorrect. The driveline should be 3 degrees or less.
Answer D is incorrect. The driveline should not be 0 degrees.

Question #132
Answer A is correct. Only Technician A is correct. A larger tank could help reduce the turbulence.
Answer B is incorrect. The inlet strainer will not reduce turbulence in the tank. It will help to control the amount that goes to the pump.
Answer C is incorrect. Only Technician A is correct.
Answer D is incorrect. Technician A is correct.

Question #133
Answer A is correct. Technician B is also correct.
Answer B is incorrect. Technician A is also correct.
Answer C is incorrect. Both Technicians are correct. The pump may have lost its prime or something could have been left in the tank restricting the flow.
Answer D is incorrect. Both Technicians are correct.

Question #134
Answer A is incorrect. Technician B is also correct.
Answer B is incorrect. Technician A is also correct.
Answer C is correct. Both Technicians are correct. The air tube should be routed away from heat sources and the line should be DOT rated.
Answer D is incorrect. Both Technicians are correct.

Question #135
Answer A is incorrect. Technician B is also correct.
Answer B is incorrect. Technician A is also correct.
Answer C is correct. Both Technicians are correct. The shaft splines could have been damaged due to the lack of grease or by the harmonics created by the engine.
Answer D is incorrect. Both Technicians are correct.

Question #136
Answer A is incorrect. The input shaft seal would not affect the flow.
Answer B is correct. Only Technician B is correct. The worn pump splines could affect the pump flow.
Answer C is incorrect. Only Technician B is correct.
Answer D is incorrect. Technician B is correct.

Question #137
Answer A is incorrect. Technician B is also correct.
Answer B is incorrect. Technician A is also correct.
Answer C is correct. Both Technicians are correct. The pulleys should be aligned with a straightedge. The pulleys in the diagram appear to be misaligned and the belt appears ready to jump off.
Answer D is incorrect. Both Technicians are correct.

Question #138
Answer A is incorrect. Technician B is also correct.
Answer B is incorrect. Technician A is also correct.
Answer C is correct. Both Technicians are correct. The air tube restriction could cause this issue or a misadjusted air regulator could be the cause.
Answer D is incorrect. Both Technicians are correct.

Question #139
Answer A is incorrect. The symbol is not a variable drive motor.
Answer B is correct. The symbol is an air motor. A hydraulic motor would have a solid triangle.
Answer C is incorrect. The symbol is not a directional check valve.
Answer D is incorrect. The symbol is not a check valve.

Question #140
Answer A is incorrect. External pump leak would not cause this issue.
Answer B is correct. Worn relief spring is the most likely cause.
Answer C is incorrect. Pump over-speeding would not cause this issue.
Answer D is incorrect. Restricted return filter would not cause this issue.

Question #141
Answer A is incorrect. Torque converter slippage would not cause this issue.
Answer B is incorrect. Constricted return filter would not cause this issue.
Answer C is correct. The pump was over-pressurized or submitted to shock loads.
Answer D is incorrect. Main relief valve, stuck in the open position, would not cause this issue.

Question #142
Answer A is incorrect. The air leak would not keep the PTO from retracting.
Answer B is incorrect. The air leak should not drain all the air from the tank if the pressure protection valve is operational.
Answer C is incorrect. Neither Technician is correct.
Answer D is correct. Neither Technician is correct.

Question #143
Answer A is incorrect. The relief valve opening too late (delayed) will not create this issue.
Answer B is correct. Only Technician B is correct. The oil temperature will rise due to the raised pressure drop in the system
Answer C is incorrect. Only Technician B is correct.
Answer D is incorrect. Technician B is correct.

Question #144
Answer A is correct. Only Technician A is correct. A larger inlet line could help reduce the chance for cavitation.
Answer B is incorrect. Lowering the tank will reduce the ability to prime the pump.
Answer C is incorrect. Only Technician A is correct.
Answer D is incorrect. Technician A is correct.

Question #145
Answer A is incorrect. Technician B is also correct.
Answer B is incorrect. Technician A is also correct.
Answer C is correct. Both Technicians are correct. The symbol is an oil tank and a well-designed tank should have a plug. Space available is important, but the system design and GPM flow will mainly affect the size of the reservoir. Ideally a tank should be two to three times the pump GPM.
Answer D is incorrect. Both Technicians are correct.

Question #146
Answer A is incorrect. Technician B is also correct.
Answer B is incorrect. Technician A is also correct.
Answer C is correct. Both Technicians are correct. An internal leak in the pump could cause loss of flow under load and cause this issue. The engine speed could also affect the pump flow.
Answer D is incorrect. Both Technicians are correct.

Glossary

Actuator A device that delivers motion in response to an electrical signal.

AH (Ampere-Hours) An older method of determining a battery's capacity.

Alternator A device that converts mechanical energy from the engine to electrical energy used to charge the battery and power various vehicle accessories.

Ammeter A device (usually part of a DMM) that is used to measure current flow in units known as amps or milliamps.

Amp Clamp An inductive-style tool that can measure the current flow in a conductor by sensing the magnetic field around it at 9 degrees Fahrenheit.

Ampere A unit for measuring electrical current; also known as amp.

Analog Signal A voltage signal that varies within a given range from high to low, including all points in between.

Analog-to-Digital Converter (A/D converter) A device that converts analog voltage signals to a digital format, located in the section of a control module called the input signal conditioner.

Analog Volt/Ohmmeter (AVOM) A test meter used for checking voltage and resistance. These are older-style meters that use a needle to indicate the values being read; should not be used with electronic circuits.

Armature The rotating component of a (1) starter or another motor, (2) generator, (3) compressor clutch.

ATA Connector American Trucking Association data link connector; the standard connector used by most manufacturers for accessing data information from various electronic systems in trucks.

Blade Fuse A type of fuse having two flat male lugs for insertion into mating female sockets.

Blower Fan A fan that pushes air through a ventilation, heater, or air-conditioning system.

CCA (Cold Cranking Amps) A common method used to specify battery capacity.

CCM (Chassis Control Module) A computer used to control various aspects of driveline operation; usually does not include any engine controls.

Circuit A complete path for electrical current to flow.

Circuit Breaker A circuit protection device used to open a circuit when current in excess of its rated capacity flows through a circuit; designed to reset, either manually or automatically.

Data Bus Data backbone of the chassis electronic system using hardware and communications protocols consistent with CAN 2.0 and SAE J-1939 standards.

Data Link A dedicated wiring circuit in the system of a vehicle used to transfer information from one or more electronic systems to a diagnostic tool, or from one module to another.

Diode An electrical one-way check valve; allows current flow in one direction but not the other.

DMM (Digital Multi-meter) A tool used for measuring circuit values such as voltage, current flow, and resistance; has a digital readout, and is recommended for measuring sensitive electronic circuits.

ECM (Electronic Control Module) Acronym for the modules that control the electronic systems on a truck; also known as ECU (electronic control unit).

Electricity The flow of electrons through various circuits, usually controlled by manual switches and senders.

Electronically Erasable Programmable Memory (EEPROM) Computer memory that enables write-to-self, logging of failure codes and strategies, and customer/proprietary data programming.

Electronics The branch of electricity where electrical circuits are monitored and controlled by a computer, the purpose of which is to allow for more efficient operation of those systems.

Electrons Negatively charged particles orbiting every atomic nucleus.

EMI (Electro-Magnetic Interference) Low-level magnetic fields that interfere with electrical/electronically controlled circuits, causing erratic outcomes.

Fault Code A code stored in computer memory to be retrieved by a technician using a diagnostic tool.

Fuse A circuit protection device designed to open a circuit when amperage that exceeds its rating flows through a circuit.

Fuse Cartridge A type of fuse having a strip of low-melting-point metal enclosed in a glass tube.

Fusible Link A short piece of wire with a special insulation designed to melt and open during an overload; installed near the power source in a vehicle to protect one or more circuits, and is usually two to four wire gauge sizes smaller than the circuit it is designed to protect.

Grounded Circuit A condition that causes current to return to the battery before reaching its intended destination; because the resistance is usually much lower than normal, excess current flows and damage to wiring or other components usually results; also known as short circuit.

Halogen Light A lamp having a small quartz/glass bulb that contains a filament surrounded by halogen gas; is contained within a larger metal reflector and lens element.

Harness and Harness Connectors The routing of wires along with termination points to allow for vehicle electrical operation.

High-Resistance Circuits Circuits that have resistance in excess of what was intended. Causes a decrease in current flow along with dimmer lights and slower motors.

Inline Fuse A fuse usually mounted in a special holder inserted somewhere into a circuit, usually near a power source.

Insulator A material, such as rubber or glass, that offers high resistance to the flow of electricity.

Integrated Circuit A solid state component containing diodes, transistors, resistors, capacitors, and other electronic components mounted on a single piece of material and capable of performing numerous functions.

IVR (Instrument Voltage Regulator) A device that regulates the voltage going to various dash gauges to a certain level to prevent inaccurate readings; usually used with bimetal type gauges.

Jump Start A term used to describe the procedure where a booster battery is used to help start a vehicle with a low or dead battery.

Jumper Wire A piece of test wire, usually with alligator clips on each end, meant to bypass sections of a circuit for testing and troubleshooting purposes.

Magnetic Switch The term usually used to describe a relay that switches power from the battery to a starter solenoid; is controlled by the start switch.

Maintenance-Free Battery A battery that does not require the addition of water during its normal service life.

Milliamp 1/1000th of an amp; 1000 milliamps = 1 amp.

Millivolt 1/1000th of a volt; 1000 millivolts = 1 volt.

Ohm A unit of electrical resistance.

Ohmmeter An instrument used to measure resistance in an electrical circuit, usually part of a DMM; power must be turned off on the electrical circuit before the ohmmeter can be connected.

Ohm's Law A basic law of electricity stating that in any electrical circuit, voltage, amperage, and resistance work together in a mathematical relationship; $E = I \times R$.

Open Circuit A circuit in which current has ceased to flow because of either accidental breakage (such as a broken wire) or intentional breakage (such as opening a switch).

Output Driver An electronic on/off switch that a computer uses to drive higher amperage outputs, such as injector solenoids.

Parallel Circuit An electrical circuit that provides two or more paths for the current to flow; each path has separate resistances (or loads) and operates independently from the other parallel path; in a parallel circuit, amperage can flow through more than one load path at a time.

Power A measure of work being done; in electrical systems, this is measured in watts, which is simply amps × volts.

Processor The brain of the processing cycle in a computer or module; performs data fetch-and-carry, data organization, logic, and arithmetic computation.

Programmable Read-Only Memory (PROM) An electronic memory component that contains program information specific to chassis application; used to qualify ROM data.

Random-Access Memory (RAM) The memory used during computer operation to store temporary information; the computer can write, read, and erase information from RAM in any order, which is why it is called random; RAM is electronically retained and therefore volatile.

Read-Only Memory (ROM) A type of memory used in computers to store information permanently.

Reference Voltage The voltage supplied to various sensors by the computer, which acts as a baseline voltage; modified by sensors to act as an input signal.

Relay An electrical switch that uses a small current to control a large one, such as a magnetic switch used in starter motor cranking circuits.

Reserve Capacity Rating The measurement of the ability of a battery to sustain a minimum vehicle electrical load in the event of a charging system failure.

Resistance The opposition to current flow in an electrical circuit; measured in units known as ohms.

Rotor (1) A part of the alternator that provides the magnetic fields necessary to generate a current flow; (2) the rotating member of an assembly.

Semiconductor A solid-state device that can function as either a conductor or an insulator depending on how its crystalline structure is arranged.

Sensing Voltage A reference voltage put out by the alternator that allows the regulator to sense and adjust the charging system output voltage.

Sensor An electrical unit used to monitor conditions in a specific circuit to report back to either a computer or a light, solenoid, etc.

Series Circuit A circuit that consists of one or more resistances connected to a voltage source so there is only one path for electrons to flow.

Series/Parallel Circuit A circuit designed so that both series and parallel combinations exist within the same circuit.

Short Circuit A condition, most often undesirable, in which two circuits, one circuit relative to ground or one circuit relative to another, connect; commonly caused by two wires rubbing together and exposing bare wires; almost always causes blown fuses and/or undesirable actions.

Signal Generators Electromagnetic devices used to count pulses produced by a reluctor or chopper wheel (such as teeth on a transmission output shaft gear), which are then translated by an ECM or gauge to display speed, rpm, etc.

Slip Rings and Brushes Components of an alternator that conduct current to the rotating rotor; most alternators have two slip rings mounted directly on the rotor shaft; they are insulated from the shaft and each other; a spring-loaded carbon brush is located on each slip ring to carry the current to and from the rotor windings.

Solenoid An electromagnet used to perform mechanical work, made with one or two coil windings wound around an iron tube; a good example is a starter solenoid, which shifts the starter drive pinion into mesh with the flywheel ring gear.

Starter (Neutral) Safety Switch A switch used to ensure that a starter is not engaged when the transmission is in gear.

Switch A device used to control current flow in a circuit; can be either manually operated or controlled by another source, such as a computer.

Transistor An electronic device that acts as a switching mechanism.

Volt A unit of electrical force or pressure.

Voltage Drop The amount of voltage lost in any particular circuit due to excessive resistance in one or more wires, conductors, etc., either leading up to or exiting from a load (e.g., starter motor); can only be checked with the circuit energized.

Voltmeter A device (usually incorporated into a DMM) used to measure voltage.

Watt A unit of electrical power, calculated by multiplying volts by amps.

Windings (1) The three separate bundles in which wires are grouped in an alternator stator; (2) the coil of wire found in a relay or other similar device; (3) that part of an electrical clutch that provides a magnetic field.

Xenon Headlights High-voltage, high-intensity headlamps that use heavy xenon gas elements.

Notes

Notes